APPLYING DATA STRUCTURES

Second Edition

APPLYING DATA STRUCTURES

Second Edition

T.G. Lewis
Oregon State University

M.Z. Smith
IBM Corporation

HOUGHTON MIFFLIN COMPANY, BOSTON
Dallas • Geneva, Illinois • Hopewell, New Jersey
Palo Alto • London

Printed in the U.S.A.

Library of Congress Catalog Card Number: 81-83273

ISBN: 0-395-31706-1

To Brian and Madeline

Green meadows
        turn to August gold.
You are the rain.

CONTENTS

PREFACE

1 WHAT IS A DATA STRUCTURE?   1
   1.1 **An Intuitive Approach**
   1.2 **Space Versus Time Tradeoff**
   1.3 **A Data Structure Classification**

2 STRINGS   15
   2.1 **Storing Strings**
   2.2 **Operations on Strings**
   2.3 **Text Editing**

3 LISTS   34
   3.1 **Dense Lists**
   3.2 **Linked Lists**
   3.3 **Multilinked Lists**

4 RESTRICTED DATA STRUCTURES   61
   4.1 **Queues**
   4.2 **Stacks**
   4.3 **Deques**

5 GRAPHS AND TREES   84
   5.1 **Definitions**
   5.2 **Storing and Representing Trees**
   5.3 **Operations on Trees**
   5.4 **Manipulating Graphs**

6 FILE STRUCTURES   122
   6.1 **Some Hardware Considerations**
   6.2 **Files**
   6.3 **File Storage and Retrieval**

7 SORTING   142
  7.1 Simple Sorts
  7.2 More Sorting Algorithms
  7.3 External Sorting

8 SEARCHING   178
  8.1 Search Effort
  8.2 Comparison of Search Algorithms
  8.3 Hashing

9 MEMORY MANAGEMENT   206
  9.1 The Fragmentation Problem
  9.2 Memory Management within an Application Environment
  9.3 The Buddy System
  9.4 Fibonacci Memory Management Systems
  9.5 A Method for Optimizing Memory Management

10 ADVANCED APPLICATIONS   224
  10.1 Sparse Arrays
  10.2 A Display Graphics Structure
  10.3 Text Editing
  10.4 Symbol Tables
  10.5 Dope Vectors
  10.6 Parsing
  10.7 I/O Processing
  10.8 Recursion
  10.9 Minimal-Path Algorithms
  10.10 B-Trees
  10.11 Extendible Hashing

APPENDIXES   275
  A String Processing with Programming Languages
  B PL/I List Processing
  C SNOBOL Data Structures
  D Pascal

GLOSSARY   316

REFERENCES   322

ANSWERS TO SELECTED EXERCISES   326

INDEX   346

PREFACE

A course in data structures is central to the study of computer science. It typically follows an introductory programming course, but is a prerequisite for advanced courses like operating systems or compiler techniques which make frequent use of the data-structure concepts and algorithms presented in this book. Representing, defining, and applying data structures is at the very heart of every computer application. The appropriate choice for storing and retrieving data may make the difference between an efficient working program and one that merely produces correct results. The structuring of data is an inherent part of solving a problem. It is both part of the problem's formulation and part of its implementation.

This second edition of *Applying Data Structures* remains an application-oriented textbook. Examples of applications show the student the usefulness and variety of data structures. The text covers the first edition's topics in more depth, but without more complexity. Enhancements include an increased number of examples and exercises. English is the language for high-level versions of the algorithms. However, where appropriate, many of the algorithms are also presented in the Pascal programming language. The Pascal implementation of selected algorithms shows how the top-down algorithms may be readily implemented in a programming language using structured programming techniques. Experienced programmers will recognize the approach, while a novice programmer will form good stylistic habits early in his or her career.

The Pascal language, unlike many programming languages, makes possible user-defined data structures. Thus it makes the implementation of many of the concepts presented here much easier and less obscure than if another language were used. (Of course, a student who does not have Pascal available may continue to program the various applications as before.) To help readers become better acquainted with Pascal, Appendix D discusses Pascal data structures. An expanded Appendix A now contains Pascal string operators as well as PL/I and SNOBOL.

In early chapters we pursue an intuitive development through examples. The book develops each topic by appealing to the reader's natural intuition.

We use mathematical formalism only where it improves clarity of presentation or gives tools to measure the performance of an algorithm. Chapter 10 deals with applications of data structures. Throughout the course one may use any section of Chapter 10, as appropriate, to reinforce concepts presented earlier in the book.

A programmer structures data in order to manipulate it. The type of structure chosen depends on the manipulation desired. Conversely, the efficiency of manipulation algorithms depends on the appropriateness of the data structure. To structure data effectively, one must know not only techniques but also when to apply them. We present basic methods of manipulating data: searching, sorting, and updating. We also give special-purpose algorithms that are linked to a particular data structure, for example, stack processing and sparse lists.

Chapter 1 presents examples that motivate the reader to study data structures. Algorithms are treated in semiformal English phrases wherever possible. We adhere to a stepwise, modular style of presentation of algorithms, reminiscent of structured programming, instead of introducing our own language for presenting them. The programming languages (Pascal, PL/I, and SNOBOL) are given only as an enrichment.

Chapter 2 starts the study of data structure by presenting unstructured data, that is, strings. We include some Pascal, PL/I, and SNOBOL to show how strings manifest themselves in a high-level language. (Appendix A and Appendix B present these topics in an expanded form.) We show the way that the string operations of concatenation, separation, substring, index, and replacement lead to problems of data organization. These problems are in turn resolved by techniques developed in Chapter 3.

Chapters 3 and 4 cover all linear data structures: lists, stacks, deques, and queues. We show that the denseness of a list is a measure of its compactness and therefore its storage efficiency. We explain that the operations of insert and delete are costly unless restricted as shown in Chapter 4.

Chapter 5 begins the formal theory of structure through graph theory. The general nature of a graph encompasses all linear and nonlinear lists. In particular, Chapter 5 models tree structures that result from restrictions placed on the graph.

Chapter 6 completes the study of structure with a presentation of data structures stored on auxiliary storage devices. Such external structures are called files. A discussion of single-key and multiple-key file structures shows that there is little difference between a file structure and a multilinked list. We do not discuss data-base organizations, other than to define a few terms and concepts. The subject of data base is too large to be dealt with thoroughly here, and is best left to a separate text.

Chapters 7, 8, and 9 cover the basic data-manipulating algorithms. Sorting is an expansive topic, so we present only the most significant techniques. Searching is also a large topic, so we discuss only the most obvious algorithms. We do, however, introduce the idea of perfect hashing: being able to locate a desired item the first time without having to examine other items during the search. Because memory management has many ramifications, we concentrate on the most frequently used methods, and present for the first time in a textbook an analytical model of memory management.

Chapter 10 contains eleven applications of data structures. These applications may be used to demonstrate a particular point or may be considered a final chapter on applications. The chapter progresses from program design to program development, and finally to program execution. Yet any application may be treated separately. This edition discusses two new topics: B^+-trees and extendible hashing. Both are techniques designed to provide fast access to records in a file.

Appendixes A, B, C, and D may be used to enrich a computer-science course that uses PL/I or Pascal for programming. Through examples, we demonstrate the constructs of Pascal and PL/I that are useful for list processing and memory allocation.

Several aspects of the book are unique. We selected the notation to accurately represent the concepts we want to reference. In Chapter 2 we use *concatenation* to mean "joining of two strings." Unfortunately, there is no comparable word that means disjoining, so we use *separation* in place of uncatenation, discatenation, and so forth.

Chapter 3 treats the concept of list density, which is a convenient way to measure storage utilization and also to name packed lists. The use of *atom* to indicate a unit of a linked list and of *NIL* for the end of a list are terms borrowed from LISP. You will note that in later chapters we prefer the terms *node* and *vertex*. Terms such as pointer, link, or thread are commonly accepted, but chain (for linked list), and shelf (for queue) are still questionable.

We dropped the contemporary usage of pre-order, post-order, and end-order for tree-search algorithms in favor of descriptive names. For instance, in place of pre-order, we use root node-left-right (NLR) order, meaning "search the root node, then its left subtree, then its right subtree." However, the other terminology is defined, so that the reader becomes knowledgeable and can read other references.

We use simple probabililty and calculus only where the use of analytical methods improves the reader's chances of understanding formulas. In general, we give performance formulas without confusing the reader with derivations that require mathematical sophistication.

For their valuable assistance, we would like to thank the following people who reviewed the first edition and helped during its developmental stages: Bob Flandrena, Bob Ford, Daniel P. Friedman, John Hamblen, May Heatherly, Sara R. Jordan, Edward Katz, David R. Musser, Ed Runnion, Victor B. Schneider, Brian Smith, Terry Walker, Udai Gupta, Elaine Weyuker, and Vivian Frederick. We also wish to thank Larry Weber for his assistance.

T. G. Lewis
M. Z. Smith

1
WHAT IS A
DATA STRUCTURE?

1.1
An Intuitive
Approach

The novice programmer is subjected to a great many details: programming language rules, the mechanics of running and debugging programs, deriving methods for solving problems and implementing them correctly. The concept of structured data and its effect on a problem's solution is rarely understood until the programmer has had some experience in doing things—and doing them correctly. For self-contained problems involving small amounts of data, the efficiency of the solution may not be as important as completing the solution rapidly. The structuring and storing of information seems unimportant. Any method that works is suitable.

This approach is no longer viable when one embarks on complex problems involving large quantities of data. The programmer must consider the problem carefully and balance the effects of structuring, storing, and retrieving data with the efficiency of the algorithm chosen to manipulate those data. Otherwise, the solution is likely to be too costly in terms of computer time and storage.

One of the first data structures programmers learn is the *array,* also known as a vector, table, or subscripted variable. An array is stored in a block of sequential memory locations. This follows the classical von Neumann model of computers that have sequentially ordered memories. An individual item may be retrieved or identified by its *address,* a unique numerical value. Programmers can reference an element in the array by using its relative position—for example, first, second, or third—together with the symbolic name of the array. The computer understands only absolute addresses (specific locations in storage). However, the language's compiler or interpreter does the conversion for the programmer.

1

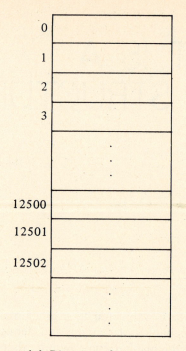

Figure 1.1 Diagram of a memory area

Figure 1.1 illustrates a computer's storage area, in which each location is a box with an associated number, the location's address. This is the way a computer sees its storage, as a block of sequential locations. Ultimately any structure is resolved into absolute addresses like these.

For now let us delay presenting the details of storage areas. Instead, let us discuss some concepts of data storage and manipulation. This follows our basic approach of considering and discussing ideas with the aid of an example before examining detailed rules and complexities. We do not want to go into so much detail that the important issues are obscured. Let us first consider some operations normally performed on data.

Suppose we have a group of people's names, and that some are listed last name first and others first name first. To make things easier, imagine that each name fits into one memory location. In reality we may need several locations per name, but would still have the problem of organization. When you examine Figure 1.2, you discover that alphabetizing the names is not easy, because you must first locate the last names.

It is easier for a person than for a computer to find the last names because a person knows that Bill, Martha, and Ann are first names. But how do we tell this to a computer? And what about the last person on the list? Is his name Sam Paul or Paul Sam? Anybody would have trouble alphabetizing this

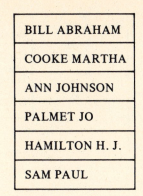

| BILL ABRAHAM |
| COOKE MARTHA |
| ANN JOHNSON |
| PALMET JO |
| HAMILTON H. J. |
| SAM PAUL |

Figure 1.2 A list of names to alphabetize

list, unless one could assume that all last names appeared first. The order or arrangement of the names is a type of structure.

Structure is a relationship between the individual elements of a group. Governments, universities, and businesses all have their defined organizations. In computer science the relationship between data items is a *data structure*. Does the idea of structure apply to some abstract quality, for example, the way we picture the data or the way we reference it? Or is it the way in which the items are stored in a computer's memory? In this book we will present various data structures, introduce their terminology, and discuss their storage. We will present numerous pertinent examples that illustrate applications of the data structures. For now, let us discuss several types of structures with which you are probably familiar, though you may not realize that they are classified as data structures.

Lists, such as the one pictured in Figure 1.2, are a type of data structure. The items are related in two ways. First, they are all names. Second, they are stored in sequential memory locations. In order to refer to the list, we need to know its beginning and either its end or its length.

Lists have some basic problems that arise when we begin to add or take away items. For example, adding to a grocery list or a client list is a simple matter of adding things at the end. If we add a name to an address list, we may try to keep the names in alphabetical order. Thus we may be forced to write in the margins or to write smaller, as shown in Figure 1.3 (next page).

Deleting items from noncomputer-type lists is equally simple. We need only to draw a line through the element. Adding names to or removing names from an unordered list may not be difficult, provided we have enough paper. However, if our list is ordered, insertion and deletion may be more of a problem. In the case of an address book, where names are added, removed, or corrected, we have limited space to make changes and to maintain the alphabetical list. In fact, ever so often we may be forced to get a new address book and start a fresh copy.

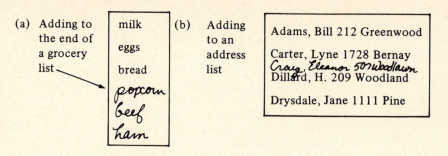

Figure 1.3 Adding to lists

A list is fairly easy to store in computer memory because in its simplest form a list is a sequence of items. It corresponds naturally to the sequential nature of computer memory. Contrasted to a list, a table (or *matrix*) is two-dimensional. As an example, consider the table in Figure 1.4, which gives prices of items purchased in units of 1, 3, or 12. Assuming that a memory location holds only one entry, the 4 × 7 entry table requires 28 locations.

Most high-level languages have built-in procedures for storing tables. To reference a single item, programmers need only give its row and column number. Thus, to find the cost of 12 units of item 1175, we reference row 3, column 4, and find 13.76. When we work in an assembly language or in a language that does not automatically store tables, we need to provide our own scheme. Usually tables are stored by column or by row (see Figure 1.5). If we know the exact size of the table (number of rows and columns), we can compute the location of any item (see Exercise 3). In a high-level language, we do not need to compute an item's location because the language's compiler or interpreter performs this computation for us.

When working with tables, we must consider such topics as inserting or deleting items, ordering the table with respect to item numbers, or searching

Item number	Price for these quantities		
	1	3	12
1062	2.99	8.84	30.50
1048	4.88	14.60	55.38
1175	1.29	3.17	13.76
1287	12.35	37.01	125.11
1296	10.12	28.92	118.05
1408	25.06	74.32	258.73
1450	7.21	20.86	80.90

Figure 1.4 A price table

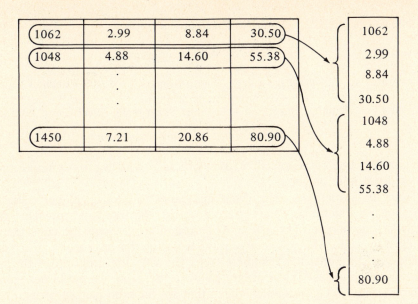

Figure 1.5 Storing a table in sequential memory locations

the table for a particular item number. These considerations present unique problems that will be discussed in later chapters of the text.

Another type of data structure is the file. A *file* is a collection of records that is usually stored outside the computer's memory on magnetic tape or disk. The common analog of a file would be the contents of a filing cabinet. The information in each folder is a *record*. Since files are usually voluminous, only portions of them can be read into memory at a single time. Their size also makes organization important so that access time will be minimized. Some files, such as the reservation records of an airline, are dynamic. People frequently make or cancel reservations and thus the file constantly changes. Other files are more or less static and require less attention; for example, census data change only every decade.

Another data structure, more complex in form, is the *tree*. Trees are frequently used to represent the organization of elements into a hierarchy. Figure 1.6 (next page) shows the order in which tasks are completed in the assembly of an airplane. The highest level (level 1) is the finished product; the lowest level (level 4) contains assembly-line tasks that may be completed concurrently. Tasks at level 4 (ribbing, sheet metal, electronics, and turbine) must be completed before tasks at level 3 (fuselage, wings, and engines) can be started. Therefore in this example we are working from the bottom to the top of the tree. Notice that the tree in Figure 1.6 is drawn upside down compared to trees found in the woods. Its branches are at the bottom and its trunk is at the top. Of course, the diagram could be turned around, but it is

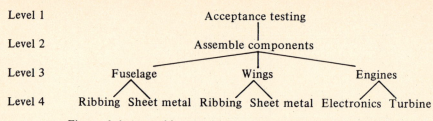

Level 1   Acceptance testing

Level 2   Assemble components

Level 3   Fuselage   Wings   Engines

Level 4   Ribbing Sheet metal Ribbing Sheet metal Electronics Turbine

Figure 1.6 Assembly tree for an airplane production line

common practice to represent a tree structure with the most important or
highest-level item on top.

Trees are sometimes used in playing games to depict a player's strategy.
In a game in which many moves are possible, the tree can represent, at the
first level, the player's first move. The second level shows the consequences
of that move and what choices are available for the next move. For simple
games, we can construct complete trees showing the progress of the game
from first play to last. Thus at any point in the game we could determine a
winning strategy. Or we could realize that we have no chance of winning.

For example, the partial game tree in Figure 1.7 shows the symmetric
moves possible at the beginning of a tic-tac-toe game. The first move could

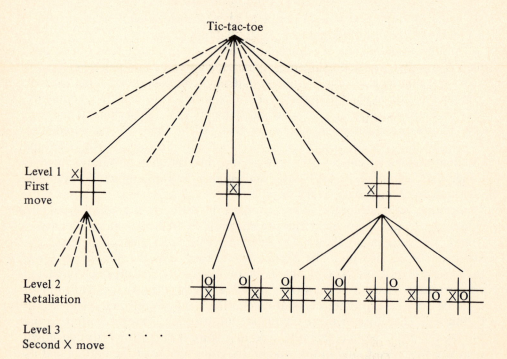

Figure 1.7 Partial tic-tac-toe game tree, showing only symmetric moves

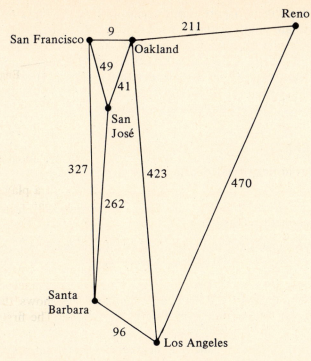

Figure 1.8 Graph showing approximate distances between selected cities

be into a corner, into the center, or into one of the side locations. Once that move is made, the retaliating move is restricted. If we finish the tree, we can see how it is possible to choose a strategy that will let us either win or tie the game. To limit the size of the tree, make the game easier to comprehend, and yet account for all possible cases, we consider only symmetric moves.

The tic-tac-toe game tree can help us develop a strategy for playing. An optimal strategy is to select a move that allows the maximum number of ways to win. Placing an X in the center square at the first move gives one four possible ways to win: having three in a row either vertically, horizontally, or with one of the two diagonals. Placing an X in a corner gives one only three ways to win; placing an X on a side, only two.

In both examples of a tree, note that a hierarchy is created. This is an important feature of trees. The hierarchy expresses either what is more important, what must be done first, or what is more general. It is a conceptualization of *single-path structures:* Each item at the lowest level may be reached by one and only one path from the item at level 1. For *multiple-path structures,* an organization called a *graph* is needed.

We can think of a graph as a road map that shows cities and the connections between them. Often if we wish to find the shortest distance between cities, we label the lengths of the connections, as shown in Figure 1.8. Graphs have wide applications in transportation to help find the shortest or fastest route that covers certain points with the least amount of backtracking. An example of this would be the routing of a school bus or a delivery truck.

At this point we will not discuss the storage of a tree or a graph in computer memory. Instead we ask you to consider this problem: How does one place a nonsequential structure, such as a tree or a graph, in sequential memory locations and accurately represent its structure?

EXERCISES

1. Define the following terms: (a) Sequential memory, (b) structure, (c) file, (d) game tree.

2. Discuss ways in which lists are allowed to grow. How does insertion differ for ordered and unordered lists?

3. Suppose that a table (such as the one in Figure 1.5) is stored in sequential memory locations. Tell how to locate any value, assuming that its row and column numbers in the table are known.

4. Devise a way for storing a graph in a computer's memory. See Figure 1.8 for an example of a graph. Consider numbering the cities so that one could use the numbers instead of their names.

5. Draw a tree showing the relationship between countries, states within the United States, and cities within each state. Where does your postal address fit into this structure?

6. List the programming languages you know and the data structures they allow. Are there structures in the language that don't fit into the categories mentioned in the text?

7. How could we organize a list of names so that it would be easy to locate all people with the same last name? Is it necessary or beneficial to separate the first names from the last?

8. Complete the rightmost branch of the tic-tac-toe game tree in Figure 1.7. How many ways are there for the X player to win? Can the O player ever win on this branch?

1.2
Space-Versus-Time Tradeoff

Behind the application of most data structures lies a space-versus-time tradeoff. Typically programmers choose between writing fast-running programs that use lots of memory and slower programs that use less memory. The choice is not always made intentionally. It may happen merely as a result of the particular implementation chosen.

For example, if we had an alphabetical list of computer users and their job account numbers, we could easily look up a particular user's number. If there were no such person, we would not need to read the entire list, since it is alphabetized. However, if we knew a job account number and wanted to find out whose it was, we would have to search the entire list. One solution to this problem would be to have two lists—one ordered alphabetically and one ordered by user number. Thus it would be easy to retrieve the information, but we would have doubled our storage requirements. If the increase were only from 100 to 200 locations, the extra storage would be negligible. But if the increase were from 10,000 to 20,000, perhaps we should consider whether we wish to sacrifice storage to gain a shorter look-up time.

Another way to make it appear that we have two lists—one in alphabetical order and the other in user number order—is to use links, as shown in Figure 1.9. The list is alphabetized, and the links tell the position the next item would have *if* the list were in order by user number.

The variable FIRST tells the location of the first user number in the list. Since FIRST = 3, the third user number in the list is the smallest. The link of 121 is 2, which says that 167 (in list position 2) is the next user number. Following 167 is 201 (in position 1), then 313 (in position 5), and finally 348 (in position 4). The link of 348 is a dash (–) which denotes the last user number. In a computer we might use zero, a negative number, or some special character to represent the end of a list. Links will be discussed in more detail in Chapter 3. This example just emphasizes the space-versus-time tradeoff. Links take up room, but give us access to the list in a different order.

The operations of adding or deleting elements from a set of data fall into the category of updating. Updating also includes correcting existing elements. Whenever we add elements, we have to consider whether or not there is room for additional data. If there is room, how do we maintain the data's structure? For example, if we have a list of numbers in descending order, how do we insert something into its proper place?

Figure 1.10 depicts the original list with location 206 shaded to show that it is not in use. To insert 1945, we must move all the numbers from 1823 to

	Name	User number	Link	
1	BAKER J	201	5	
2	CRAIG E	167	1	3
3	FARMER M	121	2	FIRST
4	JONES A	348	–	
5	WALKER T	313	4	

Figure 1.9 A list using links to show order by user number

| | | (a) Original list | 201 | 2076 | | (b) | List after inserting the value 1945 | 201 | 2076 |

(a) Original list 201 | 2076 | (b) List after inserting the value 1945 201 | 2076 |

(a) Original list 201	2076
202	1823
203	1801
204	1748
205	1607
206	//////

(b) List after inserting the value 1945 201	2076
202	1945
203	1823
204	1801
205	1748
206	1607

Figure 1.10 An ordered list (a) before and (b) after insertion

1607 down one location. Then 1945 can go into location 202 and preserve the descending order. Shifting the values down one location may not be difficult to program, but the execution time increases as we have to move more values. If the volume of data is large, we may wish to save time by keeping the original list separate from the updates and making all changes at the same time.

For example, a bank waits until the end of the day to process checks, night deposits, and banking terminal transactions. Tellers servicing customers may make individual deposits and withdrawals online, so that customers can have a record of their transactions and can learn their current balances. Thus a bank may need fast access to its customer account files during the day, but save large volumes of updates for processing after business hours. Keep this problem of updating in mind while we discuss the different structures in detail and see how difficult (or easy) it is with each structure.

A consideration arising from updating a data structure is its dynamic nature: Will the structure remain the same size or will it constantly be changing? Some languages, such as FORTRAN or COBOL, make programmers define the exact size of the arrays and tables. It does not matter if the entire area is used. Other languages, such as Pascal, ALGOL, or PL/I, allow structures to be created dynamically at execution time. This means that during execution there is a supervisory program which gets the required amount of storage for the executing program. Thus a program and its data areas may not be in contiguous memory locations (see Figure 1.11).

On the other hand, programmers can take care of their own dynamic structures. If we are using a system which requires that we request a specific number of locations, we would request some maximum amount of space based on estimates of the original size and growth potential of the data. Within that area we can let the structures grow or shrink, always being careful not to go outside the assigned area.

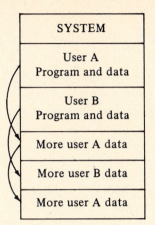

Figure 1.11 Storage and linkage of users' programs and data

Whether managed by the computer system or by programmers, dynamic structures usually require more links, tables, and bookkeeping to keep track of the current location of the data. In contrast, a static or unchanging structure may require only a pointer to designate its beginning location and a value to denote its length. Static structures lack the flexibility of dynamic structures.

We will not try here to solve any of the problems mentioned in this section. But we will discuss them in the following chapters. Keep in mind that we will discuss many different types of data structures. The major structures are lists, trees, graphs, and files. Other structures are variations of these. The choice of a data structure for a particular problem depends on the following factors:

1. Volume of data involved

2. Frequency and way in which data will be used

3. Dynamic or static nature of the data

4. Amount of storage required by the data structure

5. Time to retrieve an element

6. Ease of programming

If we have a project whose results are needed immediately, we might choose a structure that is easy to program, even though its execution time might be long. At the other extreme, we might choose a complex structure that is more difficult to program, but could be accessed in a variety of ways. Actually, there are no strict rules in choosing one particular data structure rather than another. In this book we will present the most commonly used structures, discuss their advantages and disadvantages, and give the reader a feeling for the application of the structures by providing numerous examples.

1.3
A Data Structure Classification

It may be useful in our study of data structures to remember the framework in which we will work. Figure 1.12 is a diagram of the interconnected topics discussed in this book. The classification shows how we have related the theoretical and practical aspects of applying data structures.

The emphasis of this book is on practice rather than on theory. Often, however, we rely on theory for the necessary tools to show how various methods of structuring and manipulating data compare. Figure 1.12 shows a need for graph theory, set theory, combinatorial analysis, and probability theory to analyze data structures and manipulation algorithms. Discrete mathematics is a powerful device that provides performance formulas for analysis of algorithms. Ultimately we use these formulas to select the proper algorithms.

On the practical side of our classification, we see that the study of strings, lists, trees, multilinked lists, and files leads to algorithms for manipulation of data. All applications of computers to daily problems require these structures and their algorithms.

As a demonstration, let us analyze a problem in programming. Suppose we wish to write a FORTRAN program to store the names of bank customers and their balances. In FORTRAN the only data structure that we can use is the array. An *array* is a special kind of list that we will study fully in Chapter 3.

Our problem is reduced (by FORTRAN's limitation) to working out the proper manipulation algorithms. We need an *insert algorithm* for adding new names to the bank list and a *search algorithm* for finding a given name when requested.

Now suppose we wish to theoretically analyze this problem to determine how much computer time we expect to use each time the program is run. Given that the array is n items in length (contains n names and balances), we ask: How long will it take to do an insertion? How long to do a look-up?

In Chapter 3 we show that insertion into an array requires that we move an average of $(n + 1)/2$ items. We arrive at this formula by applying probability theory and discrete mathematics.

Finally, if we know how much time t is required to move a single item in an array, we can expect it to take an average of $t(n + 1)/2$ units of time per insertion. This establishes an estimate of execution time *before* the program is written.

We selected the relationships shown in Figure 1.12 because they are representative of problems and techniques found in daily computing. We have tried to place this subset of topics in a practical application context. This means that we have sacrificed the theory half of Figure 1.12 in favor of the practical half. We urge you to read the papers and books listed in the bibliography for a theoretical perspective.

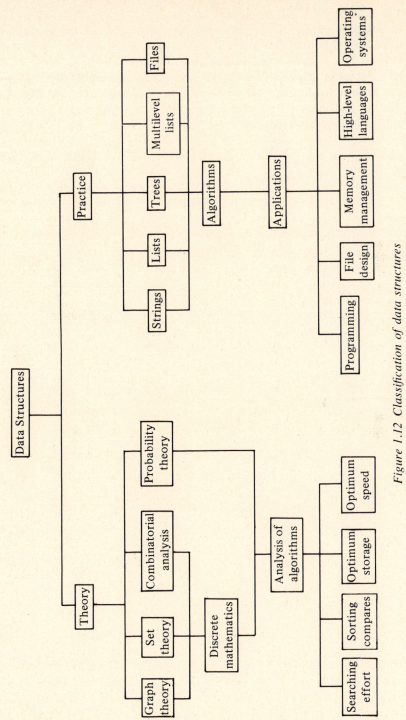
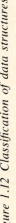

Figure 1.12 Classification of data structures

Let us now give a fuller answer to the question, "What is a data structure?" by referring to Figure 1.12. A *data structure* is an abstraction modeled by a graph, set, or combinatorial model. It is also a practical way of organizing strings, lists, trees, and so on, in computer memory.

Data structures are analyzed by theoretical means using discrete mathematics or by practical means measuring the performance of algorithms. Structures are the backbone of programming, data bases, operating systems, and other topics in computer science.

2
STRINGS

2.1
Storing
Strings

Mathematical and scientific applications prompted the design of the first computers. These computers represented numbers in binary notation, as a series of 0's and 1's, and stored the values in memory. But it was soon realized that character data could also be coded in binary notation and manipulated by computer. Thus computers became useful for business and data-processing applications.

Computers use one of two coding schemes to represent character data: ASCII or EBCDIC. ASCII is an acronym for American National Standard Code for Information Interchange. It is a standard code that simplifies the transfer of data between computers, whether by a teleprocessing network or via a portable medium such as tape. The ASCII coding uses 7 *bits* (binary digits) for each character represented. Thus the ASCII character set contains 128 different characters. Table 2.1 (pages 16–17) gives the set of ASCII characters with the codes in both binary and decimal notation.

EBCDIC (Extended Binary Coded Decimal Interchange Code) uses 8 bits for each character, thus providing for 256 unique characters. This allows room for upper- and lower-case letters, special characters, and specialized control characters that have meaning for input/output devices. Some of the 256 codes have not been assigned meanings. This allows for future expansion of the character set. Table 2.2 (pages 18–20) presents the EBCDIC coding set.

The transformation of bits into a form readable by humans occurs before the data is output, but we will not concern ourselves with the details. Instead, we will illustrate the storage of characters in a pictorial form. In Figure 2.1 (page 21), we assume that each character occupies one storage location.

binary	decimal	char	binary	decimal	char
0000 0000	0	NUL	0010 0000	32	SP
0000 0001	1	SOH	0010 0001	33	!
0000 0010	2	STX	0010 0010	34	"
0000 0011	3	ETX	0010 0011	35	#
0000 0100	4	EOT	0010 0100	36	$
0000 0101	5	ENQ	0010 0101	37	%
0000 0110	6	ACK	0010 0110	38	&
0000 0111	7	BEL	0010 0111	39	'
0000 1000	8	BS	0010 1000	40	(
0000 1001	9	HT	0010 1001	41	)
0000 1010	10	LF	0010 1010	42	*
0000 1011	11	VT	0010 1011	43	+
0000 1100	12	FF	0010 1100	44	,
0000 1101	13	CR	0010 1101	45	-
0000 1110	14	SO	0010 1110	46	.
0000 1111	15	SI	0010 1111	47	/
0001 0000	16	DLE	0011 0000	48	0
0001 0001	17	DC1	0011 0001	49	1
0001 0010	18	DC2	0011 0010	50	2
0001 0011	19	DC3	0011 0011	51	3
0001 0100	20	DC4	0011 0100	52	4
0001 0101	21	NAK	0011 0101	53	5
0001 0110	22	SYN	0011 0110	54	6
0001 0111	23	ETB	0011 0111	55	7
0001 1000	24	CAN	0011 1000	56	8
0001 1001	25	EM	0011 1001	57	9
0001 1010	26	SUB	0011 1010	58	:
0001 1011	27	ESC	0011 1011	59	;
0001 1100	28	FS	0011 1100	60	<
0001 1101	29	GS	0011 1101	61	=
0001 1110	30	RS	0011 1110	62	>
0001 1111	31	US	0011 1111	63	?

Table 2.1   ASCII Character Codes

binary	decimal	char	binary	decimal	char
0100 0000	64	@	0110 0000	96	`
0100 0001	65	A	0110 0001	97	a
0100 0010	66	B	0110 0010	98	b
0100 0011	67	C	0110 0011	99	c
0100 0100	68	D	0110 0100	100	d
0100 0101	69	E	0110 0101	101	e
0100 0110	70	F	0110 0110	102	f
0100 0111	71	G	0110 0111	103	g
0100 1000	72	H	0110 1000	104	h
0100 1001	73	I	0110 1001	105	i
0100 1010	74	J	0110 1010	106	j
0100 1011	75	K	0110 1011	107	k
0100 1100	76	L	0110 1100	108	l
0100 1101	77	M	0110 1101	109	m
0100 1110	78	N	0110 1110	110	n
0100 1111	79	O	0110 1111	111	o
0101 0000	80	P	0111 0000	112	p
0101 0001	81	Q	0111 0001	113	q
0101 0010	82	R	0111 0010	114	r
0101 0011	83	S	0111 0011	115	s
0101 0100	84	T	0111 0100	116	t
0101 0101	85	U	0111 0101	117	u
0101 0110	86	V	0111 0110	118	v
0101 0111	87	W	0111 0111	119	w
0101 1000	88	X	0111 1000	120	x
0101 1001	89	Y	0111 1001	121	y
0101 1010	90	Z	0111 1010	122	z
0101 1011	91	[	0111 1011	123	{
0101 1100	92	\	0111 1100	124	¦
0101 1101	93	]	0111 1101	125	}
0101 1110	94	¬	0111 1110	126	~
0101 1111	95	_	0111 1111	127	DEL

Table 2.1   ASCII character codes (continued)

binary	decimal	char	binary	decimal	char
0000 0000	0	NUL	0010 1001	41	
0000 0001	1	SOH	0010 1010	42	SM
0000 0010	2	STX	0010 1011	43	CU2
0000 0011	3	ETX	0010 1100	44	
0000 0100	4	PF	0010 1101	45	ENQ
0000 0101	5	HT	0010 1110	46	ACK
0000 0110	6	LC	0010 1111	47	BEL
0000 0111	7	DEL	0011 0000	48	
0000 1000	8		0011 0001	49	
0000 1001	9	RLF	0011 0010	50	SYN
0000 1010	10	SMM	0011 0011	51	
0000 1011	11	VT	0011 0100	52	PN
0000 1100	12	FF	0011 0101	53	RS
0000 1101	13	CR	0011 0110	54	UC
0000 1110	14	SO	0011 0111	55	EOT
0000 1111	15	SI	0011 1000	56	
0001 0000	16	DLE	0011 1001	57	
0001 0001	17	DC1	0011 1010	58	
0001 0010	18	DC2	0011 1011	59	CU3
0001 0011	19	TM	0011 1100	60	DC4
0001 0100	20	RES	0011 1101	61	NAK
0001 0101	21	NL	0011 1110	62	
0001 0110	22	BS	0011 1111	63	SUB
0001 0111	23	IL	0100 0000	64	Sp
0001 1000	24	CAN	0100 0001	65	
0001 1001	25	EM	0100 0010	66	
0001 1010	26	CC	0100 0011	67	
0001 1011	27	CU1	0100 0100	68	
0001 1100	28	IFS	0100 0101	69	
0001 1101	29	IGS	0100 0110	70	
0001 1110	30	IRS	0100 0111	71	
0001 1111	31	IUS	0100 1000	72	
0010 0000	32	DS	0100 1001	73	
0010 0001	33	SOS	0100 1010	74	¢
0010 0010	34	FS	0100 1011	75	.
0010 0011	35		0100 1100	76	<
0010 0100	36	BYP	0100 1101	77	(
0010 0101	37	LF	0100 1110	78	+
0010 0110	38	ETB	0100 1111	79	|
0010 0111	39	ESC	0101 0000	80	&
0010 1000	40		0101 0001	81	

Table 2.2   EBCDIC Character Codes

binary	decimal	char	binary	decimal	char
0101 0010	82		0111 1110	126	=
0101 0011	83		0111 1111	127	"
0101 0100	84		1000 0000	128	
0101 0101	85		1000 0001	129	a
0101 0110	86		1000 0010	130	b
0101 0111	87		1000 0011	131	c
0101 1000	88		1000 0100	132	d
0101 1001	89		1000 0101	133	e
0101 1010	90	!	1000 0110	134	f
0101 1011	91	$	1000 0111	135	g
0101 1100	92	*	1000 1000	136	h
0101 1101	93	)	1000 1001	137	i
0101 1110	94	;	1000 1010	138	
0101 1111	95	¬	1000 1011	139	
0110 0000	96	-	1000 1100	140	
0110 0001	97	/	1000 1101	141	
0110 0010	98		1000 1110	142	
0110 0011	99		1000 1111	143	
0110 0100	100		1001 0000	144	
0110 0101	101		1001 0001	145	j
0110 0110	102		1001 0010	146	k
0110 0111	103		1001 0011	147	l
0110 1000	104		1001 0100	148	m
0110 1001	105		1001 0101	149	n
0110 1010	106	¦	1001 0110	150	o
0110 1011	107	,	1001 0111	151	p
0110 1100	108	%	1001 1000	152	q
0110 1101	109	_	1001 1001	153	r
0110 1110	110	>	1001 1010	154	
0110 1111	111	?	1001 1011	155	
0111 0000	112		1001 1100	156	
0111 0001	113		1001 1101	157	
0111 0010	114		1001 1110	158	
0111 0011	115		1001 1111	159	
0111 0100	116		1010 0000	160	
0111 0101	117		1010 0001	161	~
0111 0110	118		1010 0010	162	s
0111 0111	119		1010 0011	163	t
0111 1000	120		1010 0100	164	u
0111 1001	121	`	1010 0101	165	v
0111 1010	122	:	1010 0110	166	w
0111 1011	123	#	1010 0111	167	x
0111 1100	124	@	1010 1000	168	y
0111 1101	125	'	1010 1001	169	z

Table 2.2   EBCDIC character codes (continued)

binary	decimal	char	binary	decimal	char
1010 1010	170		1101 0101	213	N
1010 1011	171		1101 0110	214	O
1010 1100	172		1101 0111	215	P
1010 1101	173		1101 1000	216	Q
1010 1110	174		1101 1001	217	R
1010 1111	175		1101 1010	218	
1011 0000	176		1101 1011	219	
1011 0001	177		1101 1100	220	
1011 0010	178		1101 1101	221	
1011 0011	179		1101 1110	222	
1011 0100	180		1101 1111	223	
1011 0101	181		1110 0000	224	\
1011 0110	182		1110 0001	225	
1011 0111	183		1110 0010	226	S
1011 1000	184		1110 0011	227	T
1011 1001	185		1110 0100	228	U
1011 1010	186		1110 0101	229	V
1011 1011	187		1110 0110	230	W
1011 1100	188		1110 0111	231	X
1011 1101	189		1110 1000	232	Y
1011 1110	190		1110 1001	233	Z
1011 1111	191		1110 1010	234	
1100 0000	192	{	1110 1011	235	
1100 0001	193	A	1110 1100	236	ɗ
1100 0010	194	B	1110 1101	237	
1100 0011	195	C	1110 1110	238	
1100 0100	196	D	1110 1111	239	
1100 0101	197	E	1111 0000	240	0
1100 0110	198	F	1111 0001	241	1
1100 0111	199	G	1111 0010	242	2
1100 1000	200	H	1111 0011	243	3
1100 1001	201	I	1111 0100	244	4
1100 1010	202		1111 0101	245	5
1100 1011	203		1111 0110	246	6
1100 1100	204	ɼ	1111 0111	247	7
1100 1101	205		1111 1000	248	8
1100 1110	206	ɥ	1111 1001	249	9
1100 1111	207		1111 1010	250	|
1101 0000	208	}	1111 1011	251	
1101 0001	209	J	1111 1100	252	
1101 0010	210	K	1111 1101	253	
1101 0011	211	L	1111 1110	254	
1101 0100	212	M	1111 1111	255	

Table 2.2   EBCDIC character codes (continued)

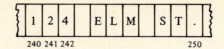

Figure 2.1 Storage of characters in memory

A *string* is a series of characters stored in a contiguous area and grouped together logically. The string may reside in a computer's memory or on an external medium such as a disk or tape. It may be input from or output to a terminal. Strings may be of any length, including zero length. A string that contains no characters is called an *empty string* (or *null string*). It has a length of zero. An empty string is not the same as a blank string, because a blank string contains blank characters and has a nonzero length.

A string can be thought of as a *list* of characters. We consider a string separately from a list because a string presents some special problems. The problems involve recognizing the end of a string, retrieving, and storing the string.

As an example of character strings, consider a short Pascal program that computes the squares of the integers from 1 to 25. Figure 2.2 shows the program written on a coding form. The squares on the coding sheet are similar to the representation of characters in memory. Each line on the coding sheet is a *record*. It contains one or more words of the program.

One method of storing the program uses a one- or two-dimensional array. We could name the array PROG. Then PROG[1] through PROG[70] would be the first record, PROG[71] through PROG[140] the second record, and so on. [*Note:* Some programming languages use parentheses instead of brackets for the subscript of an array. In this text we will use brackets.] If PROG were a two-dimensional array, the first record would be in PROG[1,1] through PROG[1,70], the second record in PROG[2,1] through PROG[2,70], and so on. Perhaps, with this example, it would be better to define PROG as an array with two subscripts. The first denotes record number and the second denotes character number within the record.

If the program were stored in memory just as shown, it would occupy 910 locations (assuming that there is one character per location). Counting all the blanks at the end of each record (that is, the number of blanks to the right of the last character on each line), we discover that there are 532. This means that more than 58% of our program storage area is blank. We omit counting the blanks that begin each line and separate the words because we assume that they were put in to make the program readable. Since 58% seems rather large, we ask ourselves if there isn't a different way.

Instead of using *fixed-length records* (records of the same length), we could consider *variable-length records* (records of varying length). At the beginning of each record, we attach a number representing the length of the record. This means that Figure 2.2 would appear as shown in Figure 2.3. The locations

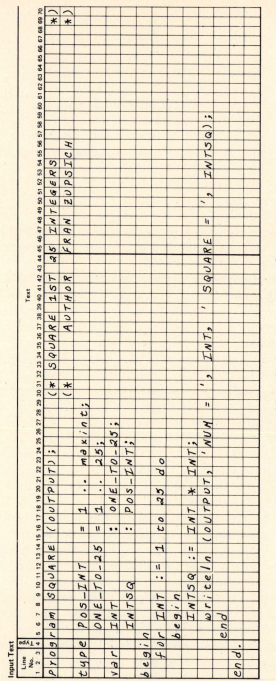

Figure 2.2 Fixed-length character strings of program text

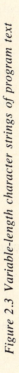

Figure 2.3 Variable-length character strings of program text

containing the lengths are outlined to make them more easily recognized. If each length value occupies one memory location, the total amount of storage required for this version is 391 locations, a considerable savings over the original 910.

Comparing the storage methods shown in Figures 2.2 and 2.3, we see that, although the method shown in Figure 2.2 uses more space, we know in advance the location of each line. The variable-length record structure requires that we refer to the record lengths. To locate any particular record, we must know where the previous record ended. This is not difficult to find if we are processing the records in sequence from first to last. But if we wish to find a middle record directly, we must add the lengths of all the preceding records.

A third strategy would be to keep a separate list of indexes to the beginning (or end) of each record. Knowing the beginning of two consecutive records, we could compute the length of the first record.

With this structure we might choose an implementation using two one-dimensional arrays. One array would contain the lengths and another array would contain the program text. If the lengths were stored in RECLEN and the program in PGMTXT, then RECLEN[1] would be 70 and the first line of the program would be in PGMTXT[1] through PGMTXT[70]. The second program line would start in PGMTXT[71].

Figure 2.4 illustrates this storage technique applied to our sample program text. Notice that there are more indexes than records. The next-to-last index locates the position following the last record. It helps locate the last character of the last record. The very last index is zero, to denote the end of the index table. Thus a program written to search for a particular record could use the zero index as a method of stopping the search.

In the next sections, we begin to deal with some of the problems associated with inserting new lines. We also discuss making changes that alter the length of one of the variable-length strings.

Where does the storage for the character strings come from? Several possibilities exist, depending on the programming language used for the character manipulation. For languages with static data structures (for example, BASIC, COBOL, or FORTRAN), the person programming has to estimate the maximum amount required for the arrays so that the storage would be set aside at compile time. For languages with dynamic data structures (for example, Pascal or PL/I), the programmer could request storage as needed during the program's execution. In this way, data areas would be only as large as necessary and not set at some predetermined maximum.

Storage management occurs at several levels in a computer system. On a large computer, the operating system has components that allocate storage to the various programs running concurrently. Chapter 9 discusses the overall management of storage by an operating system.

One level up from the operating system's storage management is the handling of storage by the program's execution time environment. The facilities

Record no.	Record indexes	Record no.	Record indexes	Record no.	Record indexes
1	1	6	222	11	309
2	71	7	247	12	368
3	141	8	252	13	375
4	170	9	279	14	379
5	195	10	284	15	0

Input Text

Line No.	Type	Text
1		program SQUARE (OUTPUT); (* SQUARE 1ST 25 INTEGERS *)
71		type POS-INT = 1 .. maxint; (* AUTHOR FRAN ZUPSICH *)
141		ONE-TO-25; INTSQ = POS-INT; ONE-TO-25 1 .. 25; var INT : be
211		gin INTSQ := INT*INT; for INT := 1 to 25 do
281		= ', INTSQ); writeln (OUTPUT, 'NUM = ', INT, ' INT SQ
351		VARE = ', INTSQ); end end.

Figure 2.4 Using indexes to locate variable-length text

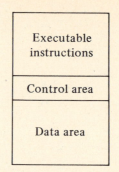

Figure 2.5 Storage diagram of a Pascal program

that are available depend on the programming language used. For example, the storage of a Pascal program looks roughly like the diagram in Figure 2.5. (An actual implementation of Pascal may have a slightly different structure, however.) There is an area reserved for variables that control execution. There is an area for executable instructions and there is an area for storage of data.

Initially the total program area comes from the operating system. But the data area is managed throughout the program's execution by instructions generated by Pascal. The instructions set aside storage for variables. The invocation of each procedure causes local variables to be defined in the data area. At the end of the procedure, the storage area can be reclaimed and reused for another procedure. If the data area ever fills up, the program can request more storage from the operating system.

EXERCISES

1. Redesign the string structure shown in Figure 2.4 so that the length of each string and the location of each string's last character are stored in END. How can we locate the first record? Compare this structure with that in Figure 2.4.

2. Define the following terms:
   (a) String, (b) substring, (c) field, (d) record, (e) variable-length record, (f) array.

3. Design an algorithm for printing the strings in Figure 2.2.

4. Write an algorithm for printing the strings in Figure 2.3.

5. Refer to the variable-length record storage for strings shown in Figure 2.4. Write an algorithm that will count the number of records in the program.

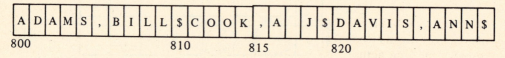

Figure 2.6 A string of names using a dollar sign ($) as a separator

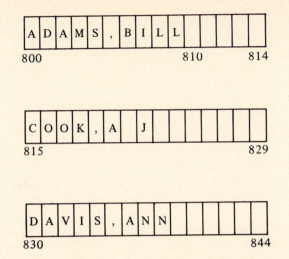

Figure 2.7 String containing substrings of 15 characters

6. Suppose that strings were stored as shown in Figure 2.6. How could we store them as shown in Figure 2.7? Compare the two storage methods. What are the advantages and disadvantages of each?

7. Find out how your computer stores strings. Is it a character-oriented machine or a word-oriented one? How many characters can it store in a "word"?

2.2
Operations
on Strings

Working with strings can be tedious if we must keep track of the exact location of each character. Some languages (SNOBOL, PL/I, and Pascal) allow the user to define and name strings. Then the user can refer to the name of the string without worrying about the way it is stored. We suggest that you do the programming exercises at the end of this section in a string-manipulation language, if possible. For more information about the string-processing capabilities of PL/I, SNOBOL, and Pascal, see Appendix A.

In this section we will refer to strings by a symbolic name rather than by a particular storage location. For example, the phrase

$$\text{TITLE} \leftarrow \text{'ONE MISTY NIGHT'}$$

denotes that TITLE is the name of the string ONE MISTY NIGHT. The arrow (←) means "has the value of," and the single quote marks (') are used to surround the actual characters in the string. The null string could be represented by two single quote marks with no space in between ('').

The most common string operations are concatenation, insertion, deletion, replacement, pattern matching, and indexing. *Concatenation* is the joining together of two or more strings to form a new string. In this book we will represent concatenation by ‖. Thus if FIRST ← 'ANN ' and LAST ← 'MARTIN', FIRST‖LAST is 'ANN MARTIN'. Notice that there is a space between the first and last name only because it was included as part of FIRST. If FNAME ← 'MARY' and LNAME ← 'COOPER', then LNAME‖','‖FNAME would be the string 'COOPER,MARY'.

Insertion and deletion are actually special cases of replacement. *Replacement* involves searching for a particular sequence of characters, removing them from the string, and inserting other characters in their place. If the empty string is inserted, we call this operation *deletion*. Remember that the empty string is *not* a string of blank characters. Therefore deletion results in a string with a shorter length than the original string. Using set notation, the ⊂ means "is a subset of" or "is included in." Therefore the statement

$$'8.00' \subset SALARY \leftarrow '8.50'$$

would be interpreted: If the string 8.00 is a substring of SALARY, the 8.00 is to be replaced by 8.50. Thus if SALARY were the string '$8.00', it would be changed to '$8.50'.

Insertion means replacing certain characters with themselves along with some other characters. For example, if SENT ← 'IF I WISH I MAY GO.', 'WISH' ⊂ SENT ← 'WISH,' will insert a comma after the word WISH.

If we want to delete the question mark from LINE where LINE ← 'HE DID IT?.', we would write '?' ⊂ LINE ← '' and the result is LINE ← 'HE DID IT.'. Notice that LINE now contains 10 characters, but it used to contain 11.

In a replacement operation the string is checked from left to right to see if it contains the specified characters. If it does, the characters are replaced. Otherwise nothing happens and the string remains the same. Sometimes we wish to know whether a string contains certain characters, but do not wish to replace them with anything. This is called *pattern matching,* and the result should be either yes or no (true or false).

After we perform insertions and deletions, the string may no longer be its original length. To verify this, we could perform a *length function,* which would tell us the number of characters in a string.

Often we may be looking not for a particular substring but for several different substrings. For example, we may wish to determine whether or not a character is alphabetic. The letter could be 'A', 'B', 'C', or any other letter of the alphabet. We will denote "or" by a vertical line (|). Thus, to see whether CHAR contains one of the letters I, J, K, L, M, or N, we would write

$$'I' \mid 'J' \mid 'K' \mid 'L' \mid 'M' \mid 'N' \subset CHAR$$

If CHAR ← 'IT' or CHAR ← '7J', the answer will be true; if CHAR ← 'OUT', the answer is false.

To copy or extract characters from a string, we perform an operation known as *substring*.[1] This operation locates consecutive characters within a string, and lets them be stored in another string. However, the substring remains part of the original. For example, suppose that we want to see if 'THE' is the first word in TITLE. We could write

if 'THE' = substring(TITLE, 1, 3) then . . .

where substring(TITLE, 1, 3) takes three characters, starting from the first character in TITLE, so that they can be compared against 'THE'. If we didn't know the location of 'THE', we might replace the 1 with a variable and make a loop searching through the string for it.

Another operation on strings is indexing. The *index* gives the position of a specified substring in the string. If the substring is not part of the string, its index would be zero. For example, if PCARD ← 'C WRITTEN BY CLARK FOSTER', the index of 'C' in PCARD is 1, since it is the first character. If there is more than one occurrence of the substring, the first substring is usually taken. Thus the index of 'C' in PCARD is 1 and not 14. The index of 'WRITE' in PCARD is zero, since that word is not part of PCARD.

Frequently when working with strings we want to break a string into its components (substrings). Compilers do this operation when scanning a line of a program to pick out key words, such as READ or WRITE, or to determine variable names and constants. We will call this operation *separation*. Let us informally discuss the way compilers make use of separation.

Example   A compiler uses the rules of a language to determine the variables and operators. For example, suppose the rule says that operators are equal signs (=) and plus signs (+), and that variables are letters and/or numbers. Then in the statement

X = A1 + B

the compiler, which begins scanning from the left, would find the variable X, the operator = , the variable A1, the operator + , and the variable B. Normally, the compiler would not include the blank following the A1 as part of the variable name. The scanning of A1 would probably proceed in the following way. Is A a letter or a number? It is a letter. So place it in a temporary location, mark it as the beginning of a variable, and go on to the next character in the string. Is 1 a letter or a number? It is a number. So place it next to A, making it part of the variable name. The next character is a blank, which we disregard and go on to the next character. Is the + a letter or a number?

[1] Some implementations of Pascal use the name "copy" instead of substring. The operation performed is the same.

It is an operator, so we have found the end of the variable. If the compiler proceeds in this manner when scanning B, it will not recognize the end of the variable until it reaches the end of the line. In some languages (for example, COBOL and SNOBOL) a blank terminates a word. Thus, when the compiler reaches the first blank after a group of characters, it has found the end of a substring.

EXERCISES

1. If A ← '', B ← 'MULE', C ← 'OLD', and D ← 'MY', what is the result of the following?

   (a) A ‖ B (b) B ‖ A (c) D ‖ C ‖ B
   (d) 'U' ⊂ B ← 'I' (e) 'C' ⊂ C (f) 'ULE' ⊂ B ← C
   (g) D ‖' '‖B (h) B ⊂ B ← (B ‖ 'S') (i) index of 'Z' in D
   (j) index of 'D' in C (k) length of A (l) length of D
   (m) substring(B, 2, 1) (n) length A ‖ B (o) length B

2. Choose a method for storing strings, describe it, and tell how you would concatenate any two strings to form a third. What would happen to the original strings?

3. Choose a method for storing strings, describe it, and tell how you would do a replacement operation. Remember that the number of characters removed is not always the same as the number of characters inserted. What will you do with the original string?

4. In the preceding example, how would you determine whether the *first* letter of a string is I, J, K, L, M, or N?

5. Using the operations described in this section, develop an algorithm that separates the units of X = A1 + B as in the preceding example.

6. Write an algorithm to search the string TITLE, 40 characters in length, to locate the first 'THE' in TITLE.

7. Modify the algorithm in Exercise 6 so that it works for strings of any length.

8. Write an algorithm to search a string named TEXT for the word 'AN'. How do you separate 'AN' from 'AND'?

9. How does BASIC (or FORTRAN, PL/I, COBOL, or Pascal) work with character strings? What operations on character strings does it allow?

2.3
Text
Editing

With the advent of mini (small) computers and CRT terminals, text editing has come into wide use. Some systems are devoted entirely to word processing. Text is entered on a keyboard, displayed on the terminal screen, edited, corrected where necessary, and finally printed in a form that looks as if it had been typed. The text editor plays an important role in this process. It is a program that allows the user to make insertions and deletions in the text, move material around, view selected portions of the text, and perform other such operations.

To simplify working with the text, the editor divides the text into lines and gives each line a number. Then, instead of saying "insert these characters," the user could say "insert this line" or "delete this line." Within each line, however, the editor works with individual characters. Thus the user can reference either an entire line (string of characters) or characters (substring) within the line. In this section we will discuss text editing and the way it utilizes the string structure. (An alternative approach to text editing appears in Section 10.3.)

On systems in which speed and fast response to the user are important, a text editor may use a fixed-length record scheme for the internal storage and manipulation of the text file. Typically the file to be edited is stored on disk, but is read into memory for editing. Then, after all changes have been made, it is written back to disk. Since the file is copied into memory, it can be stored using any scheme, so long as the file is restored to its original form when it is written out of memory. Thus a fixed-length scheme works on a variable-length file. In memory, all records become the length of the longest record. The editor uses a particular character to fill in the ends of records that are shorter than the longest. This *fill character* is usually a blank or a null character.

Using a fixed-length record structure, the text editor can begin by obtaining a large block of memory from the operating system. The text editor can set aside a small portion for control values. It allots the remainder to the text, with some space reserved for expansion of the text.

Figure 2.8 illustrates portions of the fixed-length text-editing structure. If we were to start reading the text, without studying the control information first, we would decide that the words didn't make any sense. That is because Figure 2.8 illustrates the text *after* some editing has been done. When lines are deleted, such as the line at 2161, space becomes free. Added lines can be placed in any empty place. The control information locates each line. If we look at the line number and index control information, we find that the first line is at index 2041, the second at 2081, and so on. The correct sequence in which the lines should be read is 2041, 2081, 2201, 2001, and 2121.

The allocated/free-space list, together with the beginning of free-space index, helps the text editor locate areas in which new lines can be inserted. The new lines can go either at the beginning of the free area at index 2441 or in the first free location within the text (index 2161). If new lines were always inserted at the beginning of the free area, then the beginning of the free-space area could be updated by a constant amount (the length of the record). It would not be necessary to search the allocated/free-space list. When the free-space area was exhausted and we had reached the end of our storage area, then we could reuse the deleted areas within the text.

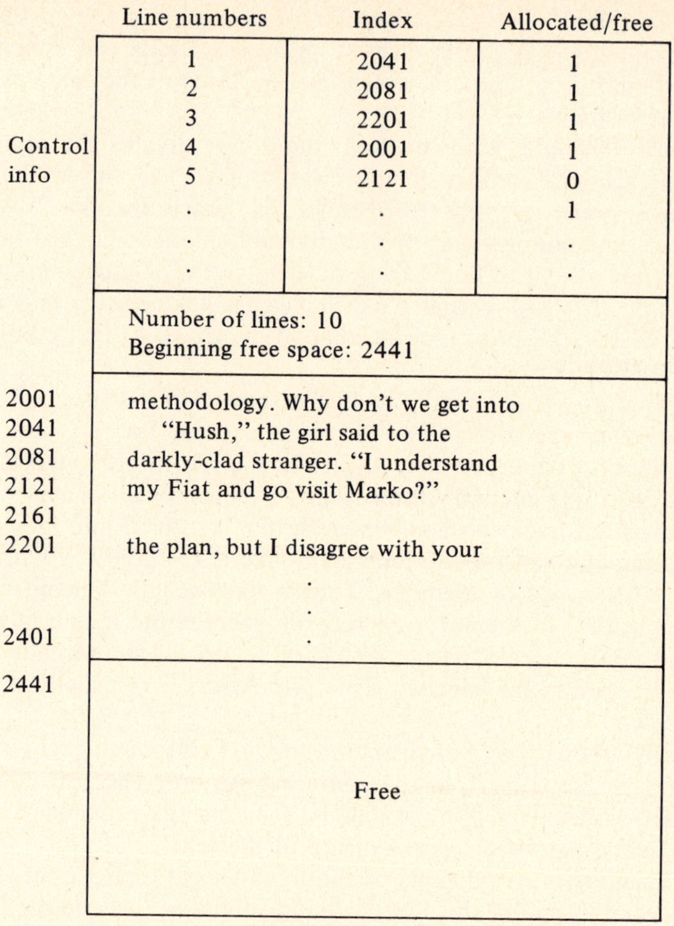
Line numbers
Index
Allocated/free
Control info
1 2041 1
2 2081 1
3 2201 1
4 2001 1
5 2121 0
1
Number of lines: 10
Beginning free space: 2441
2001 methodology. Why don't we get into
2041 "Hush," the girl said to the
2081 darkly-clad stranger. "I understand
2121 my Fiat and go visit Marko?"
2161
2201 the plan, but I disagree with your
2401
2441
Free

Figure 2.8 Fixed-length record structure for text editing

Since all records are the same length, inserting or updating a record is fairly easy. The text editor can insert a new line in any unused storage area. However, problems arise when we have to insert records into a variable-length record structure. If the new item is shorter than the free space, then our data will fit. But we must decide what to do with the leftover space. It may be too small to be of use to any other data. Then we just waste it. On the other hand, suppose that our new item is longer than any deleted record. Or suppose that an updated record is longer than its original size. Then we cannot simply replace the old data with new.

The usual solution is to place the new data in a free area, such as the one in Figure 2.8. Then, when the free area grows small, we reclaim the space left from deleted records. Management of memory space is a major problem of data structures, and we will discuss it in Chapter 9.

EXERCISES

1. What is the record length of the fixed-length records in Figure 2.8?

2. Obtain the user's manual from your computer center and read about a text editor. Which commands perform insertion, deletion, and replacement?

3. Write a program to replace 'ing' with 'e' in a text of words. Use your program on a poem of your choice. What happens to the poem?

4. Design an algorithm for moving a string from one location in memory to another. Will the algorithm work if the string in locations 213–240 is to be moved to locations 220–247? from 308–336 to 298–326?

5. Design an algorithm for printing the text in Figure 2.8.

6. Design or diagram a scheme for a variable-length storage structure for a text editor. Compare this structure with a fixed-length one.

7. What information is needed to convert a collection of fixed-length records into a collection of variable-length records?

3
LISTS

3.1
Dense
Lists

Almost every type of data structure is actually a list of some kind. Character strings are lists; arrays are lists. Files are lists that are kept on an external medium such as a disk or tape. The storage structure is often the reason for categorizing lists into various types and applying special names to them. In general, a *list* is a logical collection of atoms. An *atom* is a collection of related items called fields; and a *field* is a basic unit of information. A field may be broken further into subfields if certain manipulation is required on only a portion of a data item. When working with files, we use the term *record* more often than atom. Atom originated in the LISP programming language, developed in the 1960s.

The discussion of character strings described several methods for storing strings, which we will classify and expand on. The text-editing example in Figure 2.8 combines two types of list structures: the dense list and the linked list. Although the indexes are stored separately from the text, they are links that "link up," or provide information on how to logically connect, one text line with another and impose an order on the lines. The allocated/free list is a dense list that shows whether or not each storage record is being used.

The *dense list* is a very common and simple data structure. The word "dense" means that the list's elements are in contiguous storage locations. There are no indexes or links stored with the data, because the location of particular items can be calculated. If we examine the allocated/free list in Figure 2.8 we see that the values are 1, 1, 1, 1, 0, 1, . . . , which we interpret to mean that the first four records are allocated, the fifth is free, the sixth is allocated, and so on. In a program we could use a vector (one-dimensional array) to store the allocated/free list.

The collection of fixed-length character strings in Figure 2.2 is another example of a dense list. The characters are stored in a regular pattern; an index can locate any character. An *index* is a number that corresponds to the relative position of the item in the list. In this example we might use two indexes to locate a character—one giving line number and the other giving position of the character within the line. For example, the first character is in line 1, position 1; the second in line 1, position 2. The last character of the first line is at position 70. The t beginning line 3 is in line 3, position 1, but it is 141 actual positions from the beginning. When we use a high-level programming language, we need not bother about calculating the actual location of the data, because the compiler assigns the data to storage and generates instructions that will determine the location of the items. Before we discuss this calculation, let us study an example.

Suppose that we want to keep on a computer a telephone directory of friends and frequently called business numbers. Then, if we want to find a number, we can use the program to look it up. Our program asks for a name and responds with the number. We can arrange things so that entering a nickname or initials would give us the number and would save us the time of having to enter the entire name.

We begin the solution of our problem by deciding what information needs to be kept in our directory. Most telephone directories have name, address, and telephone number. The name is a last name, first name, and middle name (or middle initial). We will add a nickname field to make our telephone request easier. Once we decide on the data, we can estimate the size of the data fields. We will use fixed-length fields and allocate 15 characters for the last name, 15 for the first name, 1 for a middle initial, 20 for the address, 7 for the telephone number, and 8 for the nickname. If any name or address happens to be longer than the space we have allowed, we must truncate or abbreviate it. An alternative solution would be to make the fields larger, but we might be wasting a lot of storage space simply to accommodate a single name. At best, we must compromise and set aside an amount of space that is large enough for most if not all names and addresses.

Thus our telephone record or atom is 66 characters long. We can picture it as in Figure 3.1, showing the different fields. The ordering of the fields is an arbitrary one. We have placed the nickname field first, since that is probably what we will search to locate the correct atom.

Storing the atoms in sequential locations, we can think of our directory as being like that shown in Figure 3.2. Assuming that there is one character

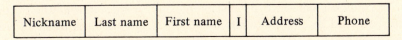

Figure 3.1 Sample telephone atom

Location

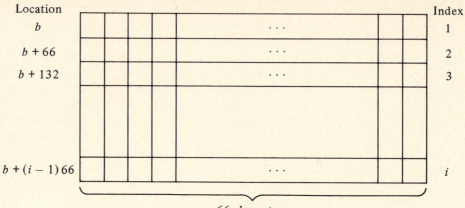

66 characters

Figure 3.2 Diagram of personal telephone directory storage

per location and that the beginning is at *b*, we know that the second atom starts at location *b* + 66, and the third one at *b* + 132. In general, the beginning of the *i*th atom (where *i* is its index) is given by the formula:

$$\text{location } i\text{th} = b + (i - 1)\ 66$$

To locate an individual field within the atom, we must add the lengths of the preceding fields. For example, the *i*th first name begins at location

$$b + (i - 1)\ 66 + 23$$

Now that we have decided on the data fields in the telephone directory, we must consider the operations that will be performed on the data. We want first to give a nickname to find out the person's telephone number. This is a *searching,* also called a table-lookup, operation. We may want to change a person's name or address (an *update* operation) or add (*insert*) or remove (*delete*) entries. Since we have already chosen a dense-list structure for the storage of our directory, we will discuss the operation of the search, update, insert, and delete activities as they apply to this type of structure. As you study these algorithms, try to determine their difficulty (that is, how easy or hard they are to implement), their speed, and their frequency of occurrence.

One other detail that is important to consider before we begin writing the algorithms has to do with whether or not the atoms are ordered in a particular way. Are they alphabetized by last name, first name, and middle initial? Or are they ordered by the nickname field? Since telephone directories are normally alphabetized, let us alphabetize ours too. As we proceed, we will discuss the implications of such a decision.

SEARCHING A DENSE LIST (Telephone Directory)

The input to our SEARCH algorithm is a nickname; the output will be the person's entire name, address, and telephone number. If the nickname is not found, an appropriate message is given. We assume that the directory is already in storage and that input and output are made via a terminal.

Although the names are alphabetized, the nicknames are not. This means that we must search the entire directory until we find the name. If the end of the directory is reached, the name is not there.

Algorithm   SEARCH

1. Comments: Search telephone directory for a particular number. Telephone directory is stored as a dense list.
    Input: NNAME   A nickname whose telephone directory entry is to be found and displayed
    Output: Directory entry, if found
                 'Nickname not in Directory', if entry not found
    Variables: NUMNAMES   Number of names in directory
                 NICKNAME[I]   The nickname field

2. Search entire directory for desired entry:
       For I = 1 to NUMNAMES
         If NNAME = NICKNAME[I] then
           Output name, address, phone number
           EXIT SEARCH

3. If no match found in step 2, then
       Output 'Nickname not in Directory'

4. End SEARCH

The "for" loop in the SEARCH algorithm says to vary I from 1 up to the value of NUMNAMES. For each value of I, the corresponding NICK-NAME[I] is checked to see if it is the one being looked for. If so, we display the directory entry and exit the SEARCH algorithm. Otherwise we repeat the loop for the next value of I. If we ever go through the entire loop without ending the search, we display the message 'Nickname not in Directory'. In an actual implementation, we would not include the test beginning step 3 because, if we arrive at step 3, we know that no match was found. The "end" in step 4 marks the end of the algorithm and implies an EXIT.

Using a dense list structure and a storage organization in which nicknames are not ordered, on the average we will have to search half the directory to locate a name. If the name is not there, we must search the entire directory. This is one of the costliest searches; but the algorithm is simple, straightforward, easy to program—and is satisfactory *if* the list is small. Alternative searching techniques are discussed in Chapter 8.

UPDATING A DENSE LIST (Telephone Directory)

Updating an entry is very similar to searching for an entry. In fact, we use the same sort of loop, looking through all the names for the appropriate one. Once it is found, we print the existing data and ask for corrections. We assume that the directory is already in storage and that a nickname specifies which entry is to be updated.

Algorithm   UPDATE

1. Comments: Update telephone directory entry. Telephone directory is stored as a dense list.
    Input: NNAME   A nickname whose telephone directory entry is to be updated

    LNAME,
    FNAME,
    MNAME,   } Data fields to replace the existing directory entry
    ADDR,
    PHON

    Output: Directory entry, if found
                 'Nickname not in Directory', if entry not found
    Variables: NUMNAMES   Number of names in directory
                 ENTRY[I]   A directory entry with subfields NICKNAME[I], LASTNAME[I], FIRSTNAME[I], MIDINITL[I], ADDRESS[I], and PHONE[I]

2. Search entire directory:
    For I = 1 to NUMNAMES
      If NNAME = NICKNAME[I], then
        Output ENTRY[I], 'Update name, address, and/or phone'
        Input NNAME, LNAME, FNAME, MNAME, ADDR, PHON
        Update directory entry:
          NICKNAME[I] ← NNAME; LASTNAME[I] ← LNAME
          FIRSTNAME[I] ← FNAME; MIDINITL[I]   ← MNAME
          ADDRESS[I]   ← ADDR;   PHONE[I]   ← PHON
        EXIT UPDATE

3. If no match found in step 2, then
        Output 'Nickname not in Directory'

4. End UPDATE

The arrow (←) in the algorithm means "is replaced by" or "takes the value of." In a programming language, it would be the replacement symbol. In PL/I and FORTRAN, it is the equal (=) sign. In Pascal it is the := symbol.

After the nickname is located in the directory, the algorithm outputs the current entry and asks the user to update the name, address, and telephone number. A more realistic and human-oriented update algorithm would probably *not* ask the user to re-enter every data field, especially if only one field had changed. Instead it might ask the user to specify which field or fields were being changed and then to input the correct information.

DELETING FROM A DENSE LIST (Telephone Directory)

When we delete an entry from the directory, we need a way of marking it so that the name will not be found. We can accomplish this by either blanking out the entire entry, making everything zero, or writing some special character in the entry to signal that it is empty. Although the SEARCH and UPDATE algorithms weren't written to handle deleted entries, with this sort of scheme, they still work. The following deletion algorithm uses the same variable names as before. It follows the same general structure as the SEARCH algorithm: It looks for the appropriate entry and then deletes it.

Algorithm   DELETE

1. Comments: Delete an entry from a telephone directory.
                The directory is stored as a dense list.
    Input: NNAME A nickname whose telephone directory entry is to be
                       deleted
    Output: 'Entry Deleted', if nickname found
                'Not in Directory', if entry not found
    Variables: NUMNAMES   Number of names in directory
                NICKNAME[I] The nickname field
                ENTRY[I]   Directory entry

2. Search entire directory:
        For I = 1 to NUMNAMES
          If NNAME = NICKNAME[I] then
            Blank entire entry: ENTRY[I] ← ' '
            Output 'Entry Deleted'
            EXIT DELETE

3. If no match found in step 2, then
        Output 'Not in Directory'

4. End DELETE

After an entry has been deleted, NUMNAMES is no longer the number of entries; it is actually the largest index of the entries. It shows the maximum size of the directory, but there is nothing to account for the number of deleted entries. If necessary, we could include something to tell the number of empty entries.

INSERTING INTO A DENSE LIST (Telephone Directory)

Inserting an entry into the telephone directory is perhaps the most difficult of the operations considered for our directory. We must solve the problem of where to put it. We have at least two choices: Add the entry at the end of the directory or add it in one of the empty spaces. If there is no empty entry, then we can insert the new one at the end and increase our NUM-NAMES counter by 1. Of course, this assumes that there is enough room to increase the size of our dense list. For now, we assume that we have allowed sufficient room for growth.

Our INSERT algorithm will try several alternatives. If it cannot find a place for the insertion, it stops with an appropriate message. The first choice for an insertion will be at the end of the current directory. Of course, this does not (or probably will not) keep our names in alphabetical order. But then, none of the algorithms took advantage of this fact, so evidently the order is not necessary. If there is no room at the end of the directory, then we proceed to search the directory for an empty entry and place the new one there. If there are no empty entries, then we must stop and report a full directory. The variable MAXSIZE tells the maximum number of entries in our telephone directory.

Algorithm   INSERT

1. Comments: Insert new entry into telephone directory. The directory is stored in a dense list.
    Input: NNAME   A nickname whose telephone directory entry is to be updated
                   LNAME,
                   FNAME,
                   MNAME,   } Data fields to replace the existing directory entry
                   ADDR,
                   PHON
    Output: Nothing if no errors made
    Variables: NUMNAMES   Number of names in directory
                   NICKNAME[I]   The nickname field
                   MAXSIZE   Maximum number of entries in dense list

2. Try to insert at directory's end:
       If NUMNAMES < MAXSIZE then
          Output 'Enter New Directory Entry'
          Input NNAME, LNAME, FNAME, MNAME, ADDR, PHON
          Increment number of directory names: NUMNAMES ← NUMNAMES + 1
          Add directory entry data:
             NICKNAME[NUMNAMES] ← NNAME;
             LASTNAME[NUMNAMES] ← LNAME
             FIRSTNAME[NUMNAMES] ← FNAME;
             MIDINITL[NUMNAMES]   ← MNAME
             ADDRESS[NUMNAMES]   ← ADDR;
             PHONE[NUMNAMES]   ← PHON
          EXIT INSERT

3. Directory has reached end of dense list area; search for an empty entry.
       For I = 1 to NUMNAMES
          If ENTRY[I] is blank, then
             Output 'Input New Directory Entry'
             Input LNAME, FNAME, MNAME, ADDR, PHON
             Add directory entry data:
                NICKNAME[I] ← NNAME; LASTNAME[I] ← LNAME
                FIRSTNAME[I] ← FNAME; MIDINITL[I] ← MNAME
                ADDRESS[I]   ← ADDR;   PHONE[I]   ← PHON
             EXIT INSERT

4. If no space in directory, then
       Output 'Directory Full, Insertion Not Made'

5. End INSERT

Most programming languages that permit the definition of a dense list require that we specify the size of the array in the program. Thus the size is fixed at compilation time. If we find during execution that our list is too small, there is no way to expand it. We must return to the original program, change it, and recompile it. The reason for this limitation is that the dense list is by definition in a contiguous area. During execution there is no way to get additional contiguous storage because something else is probably using the adjacent storage.

If we want to maintain the alphabetical order of the names in our directory, we must place our new name in the appropriate spot. For example, to include RTL, Lee, Roger, T, 550 Palm Ave., 235-1020, in our directory, our INSERT algorithm must search the names and compare Lee with each one. When it finds the one that should follow Lee, then the new entry can be inserted. If there is room for the Lee entry, then the insertion is easy. If not, then all the names following Lee must be moved down one place.

In the directory segment appearing in Figure 3.3, Lee should precede Louis and should be placed at index 29. Moving the entries around can be quite a lot of work. On the average, an insertion requires moving half the entries. Some insertions move fewer than half the entries and some move more. But the average move involves half the entries. Thus, if the directory is large, the move is significant. Later in the text we will discuss alternative structures designed to alleviate moving data to maintain order among the items.

If we are working with dense lists of numbers and not with characters, then we find that a different terminology is used. A dense list of numbers is a *vector*. The *length* of a vector is its total number of elements. For a computer, the concept of a vector or dense list is natural. Knowing the length of a vector means that the required number of memory locations can be set aside. And knowing the beginning location and index means that any element can be referenced.

Index	Nickname	Last	First	Mid	Address	Phone
.						
.						
.						
28	Steve	Landry	Stephen	G	512 Foxmoor	2232050
29	AR	Louis	Stella	R	224 Libby	2357305
.						
.						
.						
61	Leo	Morris	Allen	L	118 Tennessee	2327115

Figure 3.3 Segment of a telephone directory

Vectors are also called *one-dimensional arrays*. A dense list of vectors of the same length is called a *two-dimensional array, or matrix*. The matrix has two dimensions, *m* and *n*, where *m* is the number of vectors in the matrix and *n* is the length of each vector. A *three-dimensional array* is a collection of two-dimensional arrays, each of which has the same *m* and *n* dimensions. We can continue in this manner to define arrays of any dimension.

Logically we picture a two-dimensional array as a rectangle with rows and columns. We reference each element with two values that tell its row and its column number. For example, a 2 × 3 array A (two rows and three columns) is shown in Figure 3.4. Each box represents a single element of the array. Altogether there are six (2 × 3) elements. Physically, the array is stored in a sequential manner. Figure 3.5 contrasts the storage of the 2 × 3 array with the storage of a vector of six elements. It also shows two possible ways of storing the array: by columns as in (b) or by rows as in (c). If an array is stored by columns, we say it is stored in *column-major order,* and if by rows, *row-major order.* FORTRAN stores its arrays in column-major order; Pascal, PL/I, and COBOL store theirs in row-major order.

A[1, 1]	A[1, 2]	A[1, 3]
A[2, 1]	A[2, 2]	A[2, 3]

Figure 3.4 Logical representation of two-by-three array A

(a)	V[1]	(b)	A[1, 1]	(c)	A[1, 1]
	V[2]		A[2, 1]		A[1, 2]
	V[3]		A[1, 2]		A[1, 3]
	V[4]		A[2, 2]		A[2, 1]
	V[5]		A[1, 3]		A[2, 2]
	V[6]		A[2, 3]		A[2, 3]

Figure 3.5 Comparison of a vector V and two ways to store a two-by-three array A

A language that allows the definition of two, three, or higher multidimensional arrays stores them as dense lists. The programmer uses two (or more) subscripts to reference an element, but the computer must calculate the exact location. To do this, the high-level language compiler generates instructions that will calculate the location of each subscripted variable referenced. The compiler knows whether the array is stored by column-major or row-major order, and must also know the subscripts and the maximum dimensions of the array. For example, if the array is stored by rows, as shown in Figure 3.5(c), element A[1,2] is in the second position, A[1,3] in the third, and so on. The location of A[i,j] is $3(i - 1) + j$, where 3 is the number of columns in the array. Thus for A[1,1], $3(1 - 1) + 1 = 1$, which means that A[1,1] is in the first location. The element A[2,1] is in the fourth position, since $3(2 - 1) + 1 = 4$, and A[2,3] is in the sixth position [$3(2 - 1) + 3 = 6$]. In general, for an array in row-major order whose maximum subscripts are m and n (where m is the number of rows and n is the number of columns), the location l of the (i,j) element relative to the first element is

$$l = n(i - 1) + j$$

To compute the actual address of the [i,j] element, we add l to the location of the first element and subtract 1.

The formula for l is derived from the manner in which the array is stored. For a two-dimensional array A whose largest element is A[m,n], there are n elements in each row. If the array is stored by rows, A[1,1], A[1,2], through A[1,n] precede any element in row 2 (A[2,1], for example). Thus to find the location of an element in row 2 we must go down at least n positions in the list. A[2,1] will be in position $n + 1$; A[2,2] in position $n + 2$; and in general, A[2,i] will be in position $n + i$. To find any element in row 3 we must count down at least $2n$ elements, since there are n elements in the first row and n

elements in the second. In general, to find an element $A[i,j]$ (which is in row i, column j), we must skip over the first $i - 1$ rows which contain $(i - 1)n$ elements and add the column numbers of the element. Thus a general formula for location l of the $[i,j]$ element within the array is

$l =$ (number of rows to skip minus one) $\cdot$ (size of each row)
   $+$ column number

$l = (i - 1)n + j$

If the array is three-dimensional, the derivation is similar to that of the two-dimensional array. First we must remember that a three-dimensional array is merely a collection of two-dimensional arrays. Geometrically the three-dimensional array is a cubic configuration composed of planes. Each plane contains elements in rows and columns (see Figure 3.6). Figure 3.7 illustrates the manner in which a three-dimensional array is stored in memory by rows. This means that the first subscript has the slowest variation. All the elements with a first subscript of 1 come before those with a first subscript of 2. To find an element in the second plane, we must go down six storage locations to skip over elements in the first plane. If the largest element in the array is $B[p,m,n]$, we must skip over $m \cdot n$ elements. The product $m \cdot n$ is the size of the first plane. To locate something in the ith row of that plane, we skip over $(i - 1)n$ elements, since n tells the number of elements in each row of the plane. The complete formula for the location l of element $B[k,i,j]$, where B is stored by rows, is

$$l = (k - 1)m \cdot n + (i - 1)n + j \qquad (3.1)$$

which says that the element is in the kth plane [so we skip over $(k - 1)m \cdot n$ elements], row i (so we go down $i - 1$ rows of n elements each), and column j.

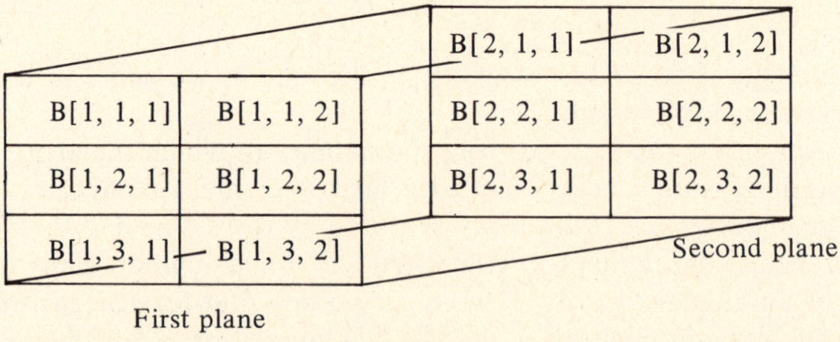

Figure 3.6 Geometric representation of a three-dimensional array

Location	Element

Location	Element			
1	B[1, 1, 1] ⎤		7	B[2, 1, 1] ⎤
2	B[1, 1, 2] ⎥		8	B[2, 1, 2] ⎥
3	B[1, 2, 1] ⎥ First plane		9	B[2, 2, 1] ⎥ Second plane
4	B[1, 2, 2] ⎥		10	B[2, 2, 2] ⎥
5	B[1, 3, 1] ⎥		11	B[2, 3, 1] ⎥
6	B[1, 3, 2] ⎦		12	B[2, 3, 2] ⎦

Figure 3.7 Storing a three-dimensional array by rows

Example Compute the location of B[1,2,1]. From the example in Figure 3.7 we see that the answer should be 3. The largest element of B is B[2,3,2], so $p = 2$, $m = 3$, and $n = 2$. Also, $k = 1$, $i = 2$, and $j = 1$.

$$l = (k - 1)m \cdot n + (i - 1)n + j = (1 - 1)3 \cdot 2 + (2 - 1)2 + 1$$

$$= 0 \cdot 6 + 1 \cdot 2 + 1 = 3 .$$

A dense list, whether it is a vector or an array, contains no pointers to specific elements, because an element's location can be computed relative to the beginning of the list. We must remember only the appropriate formula and the location of the first element. A structure that contains pointers or indexes, as in Figure 2.8, takes additional memory for the pointers. The function of a pointer is to assist in handling information. Therefore, to utilize memory efficiently, it is desirable to minimize the size and number of pointer fields.

A measure of a structure's utilization of memory is its density. *Density* is the ratio of information to the total space required. Information and total space may be measured in number of characters, number of words, or number of bits, so long as both are described in the same unit of measure.

$$\text{Density} = \frac{\text{information size}}{\text{total space}}$$

A dense list has density 1.0. The total number of characters exactly equals the number of information characters; there are no pointers. A dense list structure is the best utilization of memory. Why then do we need other structures?

A dense list is useful for data that does not change frequently. When we want to add a new name to the telephone directory, we must move atoms to insert the new entry. We stated previously that this means moving an average of half the directory. Let us see how we arrive at that conclusion. Suppose that we have a list with n entries and wish to insert an atom at position k. Then $n - k + 1$ moves are necessary. For example, to insert the value 24 into the list in Figure 3.8, the atoms with indexes 4 through 8 must be moved

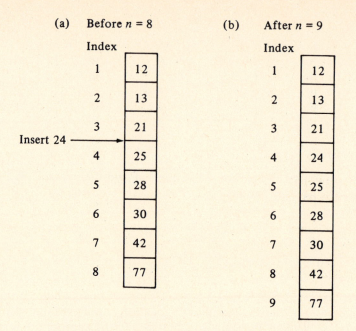

(a)   Before $n = 8$ (b)   After $n = 9$

Figure 3.8 Insertion of 24 into index 4 of a list

into positions 5 through 9. Then 24 is inserted and has index 4. We assume that movement takes place at the end of the list. After an insertion, the value of n increases by 1, which means that an insertion may cause an *overflow* if the list exceeds the available space.

The average number of moves made during an insertion depends not only on the number of items to move but also on the probability that the insertion will take place at a particular location. If the operation occurs near the end of the list, fewer elements need to be moved than if the operation occurs near the beginning. However, if most insertions occur near the beginning, the average will be close to n, n being the total number of elements in the list. In general, the average number of moves during an insertion (m_i) is given by the equation

$$\text{Insertion: } m_i = \sum_{k=1}^{n} (n - k + 1) \cdot p_k$$

where p_k is the probability of inserting an atom into location k.

The probability of accessing any atom in an insertion, update, search, or deletion operation depends on many factors, such as the data in the list, the reason for updating an item, and so on. Unless we know otherwise, we may reasonably assume that every atom is equally likely to be accessed. That is, the probability of accessing any atom out of a total of n is $1/n$ ($p_k = 1/n$ for all k). Then we may compute m_i:

$$m_i = \sum_{k=1}^{n} (n - k + 1)\frac{1}{n} = \frac{1}{n} \sum_{k=1}^{n} (n - k + 1)$$

$$= \frac{1}{n} \left[\sum_{k=1}^{n} (n + 1) - \sum_{k=1}^{n} k \right]$$

$$= \frac{n(n + 1)}{n} - \frac{1}{n} \cdot \frac{n(n + 1)}{2} = n + 1 - \frac{n + 1}{2}$$

$$= \frac{n + 1}{2}$$

On the average, half the list must be moved during an insertion. If the list is long, we are paying a high cost to update a dense list structure.

Example If a dense list contains 500,000 atoms, what is the average number of moves to make during an insertion?

$$m_i \approx \frac{500,000}{2} = 250,000$$

(The $\approx$ symbol means "is approximately equal to.") If our computer is fast enough to move an atom every 20 microseconds (20×10^{-6} second), how long will the average insert take?

$$\text{Move time} \approx (20 \times 10^{-6} \text{ second}) \cdot (250,000) = 5 \text{ seconds}$$

If the typical insert operation involves 1% of the atoms in the list, how much computer time is required to perform the update?

$$\text{Update time} = (.01)(500,000)(5 \text{ seconds}) \approx 7 \text{ hours}$$

The previous example shows how costly a dense list structure may be even though it utilizes memory efficiently. In Section 3.2 we will see how the use of pointers can eliminate the necessity to move information around in memory.

EXERCISES

1. If a list contains 32 atoms, how many bits are needed to point to any one of the atoms? In other words, how many bits long must the index be? (*Hint:* A vector of length 128 needs an index of seven bits.) How many bits are needed for the index of a list whose length is 8192; whose length is 10,000?

2. Define the following terms:
   (a) List, (b) dense list, (c) density, (d) index, (e) vector, (f) array, (g) subscript, (h) matrix.

3. Derive a formula for the location (index) of an element of a two-dimensional array C that is stored in column-major order. The maximum subscripts of C are C[4, 5]. Sketch a picture of the order in which the elements of C are placed in memory.

4. Derive a formula for the location of an element of a three-dimensional array X stored in column-major order. The maximum subscripts of X are X[s_1, s_2, s_3]. Explain the components of the formula.

5. In general, the correspondence between the subscripts [j_1, j_2, . . . , j_n] of an n-dimensional array stored in row-major order is given by the following formula. (The maximum subscripts of the array are A[d_1, d_2, . . . , d_n].)

$$\text{Location} = \sum_{r=1}^{n-1} (j_r - 1) \prod_{s=r}^{n-1} d_{s+1} + j_n$$

First, assume that $n = 3$ and show that this formula produces the one in the text (3.1) if p is substituted for d_1, m for d_2, n for d_3, and so on. Then explain in general terms the components of the formula.

$$\textit{Note:} \prod_{s=r}^{n-1} d_{s+1} = d_{r+1} \cdot d_{r+2} \cdot d_{r+3} \cdots d_n \quad \text{(pi product)}$$

6. Derive a formula for the location (index) of the element of an array stored in row-major order whose subscripts may be positive, negative, or zero. The declaration of the array is A[l_1:u_1, l_2:u_2, l_3:u_3, . . . , l_n:u_n], where l_i is the lowest value of the ith subscript, and u_i is the largest value of the ith subscript. For example, A[2:3, −3:0] defines an array with eight elements A[2, −3] through A[3, 0] (see Exercise 5).

7. What are the assumptions involved in computing the average number of moves made to update a dense list?

8. Calculate the location of A[3, 6, 4] of an array in row-major order whose maximum subscript is A[10, 20, 5]. What would its position be if the maximum subscripts were not known?

9. What problems are encountered in looking up a person's name whose telephone number is known? What if the number is not in the directory?

10. Write a computer program that will input the values of n, d_1, d_2, . . . , d_n and compute the location of any element A[j_1, j_2, . . . , j_n]. Use the formula in Exercise 5.

11. Write a program to implement one of the telephone-directory algorithms.

12. Modify the INSERT algorithm to maintain alphabetical order of the names.

13. Assume that nicknames in the telephone directory are kept in alphabetical order. Then rewrite the telephone-directory SEARCH or UPDATE algorithm so that it takes advantage of the order of the nicknames. Is an entry found any faster? What if the nickname is not in the directory?

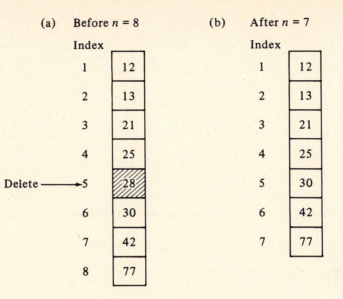

Figure 3.9 Deletion of atom 5 from a list

14. Describe the telephone-directory data items in a programming language of your choice. Do the SEARCH, UPDATE, INSERT, and DELETE algorithms require modifications so that they will work with your description?

15. Suppose that a delete operation on a dense list does not create empty spaces, but instead moves entries up to keep the list in contiguous locations. (Use Figure 3.9 as an example.) Compute the average number of moves required during a delete.

3.2
Linked
Lists

Although the dense list is a very simple structure, it does have some disadvantages. It is difficult to make insertions or deletions if the list must be kept in order. Another problem arises if we try to change the size or shape of the list. Suppose, for example, that we have a list that occupies 200 locations. That is, somewhere in memory there are 200 contiguous locations that we have set aside for the list. What if we wish to double the size of the list? In most cases we cannot use the 200 locations that follow the list, because they probably are reserved for other data. We cannot use just any 200 locations because our list must be contiguous. In some programming languages (FORTRAN and COBOL), expanding a list means recompiling the program to allocate the exact number of requested locations. Other languages (ALGOL, APL, Pascal, and PL/I) allow the programmer to change the size of lists or arrays during execution. The processor of the language reserves a large block of memory for defining or redefining data structures as requested by the programmer.

To make a list more flexible and easier to update, we can include a pointer (link) with each atom in the list. In a *linked list* each atom contains a pointer that tells the location of the next atom. The pointer eliminates the need to store the atoms of a list in a contiguous memory region. To illustrate how pointers are used to logically connect the elements in a list, we will consider a portion of the telephone directory. We will begin with a large block of memory and assume that each atom occupies 60 locations (29 for name, 20 for address, 7 for telephone number, and 4 for the pointer). The atoms and pointers are initially in the same order (see Figure 3.10). Our structure is basically a dense list with a pointer field included. The hyphen (-) in the last pointer field represents the end of the list. It is called the *nil pointer* because it points to nil, or nothing. In the computer we would use a negative number, zero, or anything that is not a valid pointer. The first atom in the list is special; we will discuss it shortly.

Let us first add a name to the directory. We can place it at the end of the list (beginning in location 2360) and change pointers so it is in the correct alphabetical position. For example, suppose that the entry is Lang, Al, 311 Moss, 236-1111. To decide where the name should go, we begin with the first name, which is Abel. Since Lang does not come before Abel, we look at Abel's pointer to see which name comes next. The next name is in the next set of memory locations 2060–2120. Since Lang does not come before Baker, we use Baker's pointer to go on to Carter. Lang does not come before Carter, but on going to the next name we find that Lang does belong before Minte. This means that Lang's pointer should be 2180. Now we must change Carter's pointer so that it leads to Lang. The updated pointers and inserted name

Location	Information					Pointer
2000	Abel,	J.	G.	110 Oakleaf	236-4010	2060
2060	Baker,	Sue		409 Sunset	784-1182	2120
2120	Carter,	L.	H.	17 Bernay	785-1365	2180
2180	Minte,	Al		204 Pine	236-7295	2240
2240	Pont,	M.	R.	1 Market	480-1027	2300
2300	Sands,	T.	H.	671 First	784-8240	—

Figure 3.10 Telephone directory with pointers

Location	Information					Pointer
2000	Abel,	J.	G.	110 Oakleaf	236-4010	2060
2060	Baker,	Sue		409 Sunset	784-1182	2120
2120	Carter,	L.	H.	17 Bernay	785-1365	2360
2180	Minte,	Al		204 Pine	236-7295	2240
2240	Pont,	M.	R.	1 Market	480-1027	2300
2300	Sands,	T.	H.	671 First	784-8240	–
2360	Lang,	Al		311 Moss	236-1111	2180

Figure 3.11 Telephone directory after inserting Lang

appear in Figure 3.11. Note that the list no longer appears in alphabetical order, as it does in Figure 3.10. However, the pointers keep the order intact.

Let us now delete Al Minte from the list. We must first locate his name and find the atom that points to him. Currently, Al Lang's pointer (2180) references Al Minte, and Al Minte's pointer is 2240. To delete Minte, we must change Lang's pointer to 2240; this is demonstrated in Figure 3.12. Even though Minte's name appears in the list, there is no pointer to it and locations 2180–2239 are available for an insertion. We must now mark these locations so that they may be used later. We could keep a list of free locations or we

Location	Information					Pointer
2000	Abel,	J.	G.	110 Oakleaf	236-4010	2060
2060	Baker,	Sue		409 Sunset	784-1182	2120
2120	Carter,	L.	H.	17 Bernay	785-1365	2360
2180	~~Minte,~~	~~Al~~		~~204 Pine~~	~~236-7295~~	~~2240~~
2240	Pont,	M.	R.	1 Market	480-1027	2300
2300	Sands,	T.	H.	671 First	784-8240	–
2360	Lang,	Al		311 Moss	236-1111	2240

Figure 3.12 Telephone directory after deleting Minte

could make the phone number 000-0000. We will discuss ways to recover unused memory locations in Chapter 9.

The last atom in a linked list has a nil pointer to show that no atom follows it, so it is frequently called the *tail*. The first atom is the *head* of the list and is pointed to by a separate atom called the HEADER. The HEADER may contain additional information about the list, such as name, creation date, length, or security information to limit access to the data. The lists in Figures 3.10 through 3.12 should contain the following HEADER.

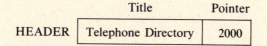

If insertions or deletions occur at the head of the list, we must change the HEADER pointer. If the entire list is deleted, the HEADER will be retained. Its pointer field will be nil, but as long as the HEADER is there the list is still available for use.

Let us describe how pointers generally change during an insertion or deletion. We assume that $atom_i$ has $pointer_i$ and is logically followed by $atom_{i+1}$ with $pointer_{i+1}$. In other words, $pointer_i$ tells the location of $atom_{i+1}$. To insert $atom_{new}$ between $atom_i$ and $atom_{i+1}$, we must do the following. We use the HEADER atom to locate the beginning of the linked list. Then we follow the pointers until $atom_i$ is reached. At this point the LINK_INSERT algorithm below is performed.

Algorithm   LINK_INSERT

1. Comments: Basic part of algorithm to insert an atom into a linked list. This assumes that the atom goes between $pointer_i$ and $pointer_{i+1}$.

2. Get storage for $atom_{new}$:
    newstor ← storage location for $atom_{new}$
    Place data for $atom_{new}$ beginning in location newstor

3. Adjust pointers, thus inserting atom:
    $pointer_{new}$ ← $pointer_i$
    $pointer_i$ ← newstor

4. End LINK_INSERT

We begin the algorithm by obtaining storage for our new atom. In Pascal we would use the NEW function, which would reserve enough storage for the atom's data and would return the location of that storage. We save that location's value in newstor. Step 2 in the insertion algorithm copies the new atom's data into the storage just obtained. The next two operations in step 3 are to adjust the pointers so that the new atom is in the correct order. Figure

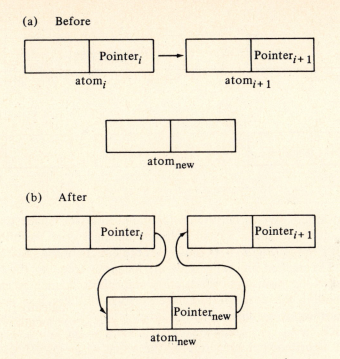

(a) Before

atom$_i$ atom$_{i+1}$

atom$_{new}$

(b) After

atom$_{new}$

Figure 3.13 Insertion of atom$_{new}$ between atom$_i$ and atom$_{i+1}$

3.13 graphically illustrates the states before and after the LINK_INSERT algorithm has taken place.

Deletions are handled in a similar manner. To delete atom$_i$ we locate atom$_i$ and its predecessor atom$_{i-1}$ and perform the DELETE_LINK algorithm.

Algorithm DELETE_LINK

1. Comments: Basic part of linked-list deletion algorithm. Algorithm assumes that atom$_i$ (the atom to be deleted) has been located and that its predecessor is atom$_{i-1}$. That is, we are not deleting the head of the list.
    Input: pointer$_i$ Pointer to atom$_i$, the atom to be deleted
                 pointer$_{i-1}$ Pointer to atom$_{i-1}$, the atom preceding atom$_i$
    Output: Nothing

2. Point to atom$_{i+1}$:
        pointer$_{i-1}$ ← pointer$_i$

3. Release storage space for atom$_i$

4. End DELETE_LINK

We delete the atom merely by changing pointer$_{i-1}$ so that it points to atom$_{i+1}$. Then we release the space formerly occupied by atom$_i$. In Pascal

(a) Before

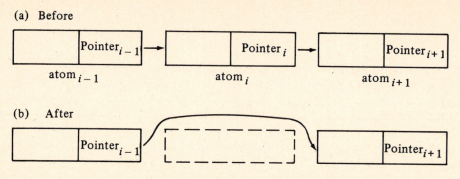

Figure 3.14 Deletion of atom$_i$

this is performed by the DISPOSE function. Released space may later be reused by other variables. Pascal's storage manager does the allocation and releasing of storage. Figure 3.14 illustrates the effects of the DELETE_LINK algorithm. The dashed line represents the locations freed by the deletion.

The linked-list insertion and deletion algorithms do not include specific steps for finding the desired atom. We perform this function in a separate FIND algorithm. It assumes that the atom to be located is identified by a particular KEYVALUE, which must match atom$_{key}$—the key field of the atom. The FIND algorithm not only returns the value of ATOMPTR, the pointer to the found atom, it also returns PREVPTR, the pointer to the preceding atom. The PREVPTR is used by the linked-list DELETE_LINK algorithm, where pointer$_{i-1}$ = PREVPTR and pointer$_i$ = ATOMPTR.

Algorithm   FIND

1. Comments: Find atom in linked list. The atom to be found has a key equal to KEYVALUE. This algorithm will not work if the list is empty.

    Input: KEYVALUE   A key used to locate a particular atom in a linked list

    Output: ATOMPTR   A pointer to the atom with key equal to KEYVALUE

    PREVPTR   A pointer to atom preceding one with key equal to KEYVALUE

    Variables: HEADER   A pointer to head of the linked list

2. Initially there is no preceding atom:
    PREVPTR ← NIL

3. Point to first atom in linked list:
    ATOMPTR ← HEADER pointer

4. Search linked list at least once; stop when PREVPTR is NIL:
    If KEYVALUE = atom$_{key}$ at location ATOMPTR, then
    EXIT FIND
    Else
    PREVPTR ← ATOMPTR
    ATOMPTR ← pointer field of atom at location ATOMPTR

5. Atom not found; PREVPTR and ATOMPTR are NIL

6. End FIND

The linked-list FIND algorithm works if the list has one or more entries, but it has no check to see if the list is empty. An implementation of the algorithm should insert this test somewhere before step 4. The operations inside the step 4 loop will be performed at least once. The test to see if PREVPTR is NIL (that is, that we have reached the end of the list) is made at the end of the loop. If the atom has been found, we exit from FIND. If not, we set PREVPTR and ATOMPTR to point to the next atoms in sequence and repeat our search.

The linked-list DELETE and INSERT algorithms will not work if the deletion or insertion is to be made at the beginning of the list. In this special case we must update the HEADER pointer also. These refinements to DELETE and INSERT are left as exercises for you to do (see Exercises 4 and 8 at the end of this section).

A linked list requires more storage and more complex algorithms than a dense list. For example, suppose that the linked-list representation of the telephone directory requires 29 characters for the name, 20 for the address, 7 for the telephone number, and 4 for the pointer. We can compute the density as

$$\text{Density} = \frac{29 + 20 + 7}{29 + 20 + 7 + 4} = .93$$

The density of the dense list is 1.0, so the linked list uses more storage. However, update operations do not require us to move any atoms. At most we need to change two pointers.

We have not yet considered the amount of time required to search a list for an item that should be deleted, or to locate the place where an item should be inserted. On the average, assuming that the probability of updating any atom is equally likely, we should expect to read through half the list. Whether we use a dense list or a linked list structure, we will have to include a search as part of the update operation. (Searching is discussed in Chapter 8.) The real advantage of a linked list is that we need not move the atoms around the way we must with a dense list.

EXERCISES

1. Why are linked lists needed and used in spite of their cost in memory utilization?

2. What is the purpose of the HEADER atom? What happens if there is no HEADER and the list is empty?

3. How do we determine the number of bits needed in the pointer field of an atom?

4. Modify the linked-list INSERT algorithm so that it uses FIND and will allow insertions at the beginning of the list. Should it check for an empty list?

5. Given a telephone directory as described in the text and a telephone number, write an algorithm that will produce the name of the person belonging to that number. If there is no such number in the directory, write a message to that effect.

6. Assuming that it is equally likely to access any atom in a linked list, show that on the average we must examine $n/2$ atoms before finding the one we are seeking. (The total number of atoms is n.)

7. Write a program that will store and update a small telephone directory in memory in the form of a linked list. Print the original directory. Then make some insertions and deletions and print the final directory. Use any format for input and output you desire.

8. Modify the linked list DELETE algorithm so that it uses FIND and will allow deletions at the beginning of the list. Does anything special have to be included to delete the last item in the list? What if the list contains one item and that item is being deleted?

9. Define a set of data (character or numeric) and implement the FIND and INSERT algorithms. To check your program you may want to write a PRINT routine to print the linked list.

10. Use a programming language of your choice and tell how it manages storage for your programs at execution time. If it does not have functions like NEW and DISPOSE, how can your program manage its own storage? How expensive are the storage management functions to use? Will they slow down the execution of your program?

3.3
Multilinked Lists

Often a single pointer for each atom is not enough. In the deletion of an entry from a linked list, we had to locate not only the item to delete but also its predecessor. If every atom had two pointers instead of only one, we could use one to locate the previous atom and the second to locate the following atom. This would simplify our deletion and find algorithms.

Another use for more than one pointer is to keep track of a different order on the list. For example, in the telephone directory, the pointers ordered the names alphabetically. How would we tackle the problem of finding a person's name if given only the number? Searching would not be an easy task. We would have to read every entry in the directory just to discover that the number is not listed.

Lest we run the telephone directory into the ground, let us discuss a similar problem in a different field. The First National Bank of Boulder Creek has numerous accounts. It is a small-town friendly bank that wants to keep its

Index	Name	Alpha pointer	Account	Account pointer
1	ADAMS, D. A.		9089	
2	BAKER, J. R.		1195	
3	CARR, D. M.		1034	
4	DAVIS, J. H.		3782	
5	MILLER, I. M.		6217	

Figure 3.15 Initial listing of customer accounts

customers happy by using the customer's name. However, the bookkeeping department must deal with the Federal Reserve Bank, which works entirely with account numbers. Therefore they need a way to provide easy access to the accounts—either by account number or by customer name. A solution to this problem is to use a *multilinked list,* a list in which each atom has two or more pointers. Let us see how pointers with each atom can provide the type of data structure that this bank needs.

Let us suppose initially that the accounts have been alphabetized and stored in memory. To simplify matters, we will refer to the atoms by index rather than by memory location and will make the account numbers four digits. A sample listing of the accounts appears in Figure 3.15. We will first discuss the two pointer fields: alpha pointers and account pointers.

The alpha pointers will show the alphabetical listing of customers. Initially, each alpha pointer is one more than the index of the atom, since the names are in alphabetical order. After several insertions or deletions this may no longer be true. The account pointers are more difficult to define, except in this short example where we could do it by inspection. If the list were longer, it would be almost impossible to do by hand and could be accomplished with a sorting program. However, since each account has an index to tell its location, we can use the index and account numbers to determine the account pointers.

We will first order the accounts numerically, keeping the index with each account number. The result is shown in Figure 3.16. From it we can determine the account pointers. Each account should be linked to the one following it,

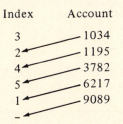

Index   Account

Figure 3.16 Accounts with index listed in ascending order

as represented by the arrows. Account 9089 has no successor, so its pointer will be nil. Figure 3.17 shows the complete listing of customer accounts with alpha and account pointers. The two HEADERS show the first element in each ordering.

Insertions and deletions require more effort with the multilinked list than with the singly linked list because the former has more pointers to change (see Exercises 1 and 2 at the end of this section). The multilinked list also requires more memory for pointers. However, pointers provide an alternative to having two copies of the lists, each in a different order. And we can work with the accounts in either alphabetical or numerical order.

A *doubly linked list* is a multilinked list in which two pointers are used to keep track of an atom's predecessor as well as its successor. A *forward pointer* shows the atom that follows and a *backward pointer* shows the predecessor. Figure 3.18 shows the representation of a doubly linked list. In general, the forward pointer of $atom_{i-1}$ is the same as the backward pointer of $atom_{i+1}$; that is, they both point to $atom_i$. The first atom has a nil backward pointer and the last atom has a nil forward pointer. The HEADER points to the first atom and the TAIL points to the last.

Throughout our discussion all the lists have been *linear,* with a beginning (first) element and an end (last) element. An alternative kind of list, whether it has single or multiple links, is the *circular list* or *ring.* In this structure the last element points to the first and there is no nil pointer (see Figure 3.19). Often a circular list is used to implement a work area or buffer for input and output. We store data in the first area (or first few areas) and can continue storing data until we are back at the beginning of the list. Concurrently, while we are placing data at the end, we can remove other data from the beginning. The circular nature of the ring thus allows the same memory area to be reused. Of course we must be careful not to let the ring overflow by writing new data over old data before the old data can be used.

To visualize the nature of the circular list, consider an automatic bottling machine. The empty bottles approach the machine on a linear conveyor belt.

Alpha header	1		Account header	3	
Index	Name	Alpha pointer	Account	Account pointer	
1	ADAMS, D. A.	2	9089	–	
2	BAKER, J. R.	3	1195	4	
3	CARR, D. M.	4	1034	2	
4	DAVIS, J. H.	5	3782	5	
5	MILLER, I. M.	–	6217	1	

Figure 3.17 Final listing of customer accounts

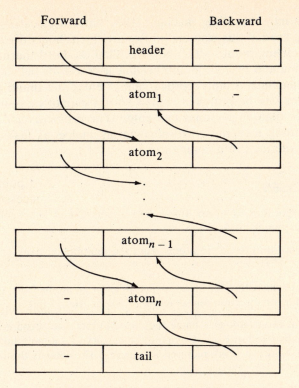

Figure 3.18 Doubly linked list

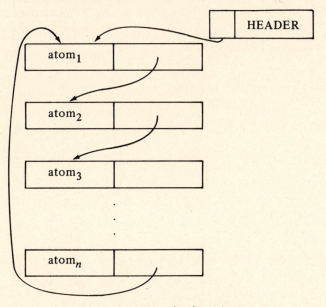

Figure 3.19 Circular list (ring)

As they pass into the filling machine, they enter a circular conveyor belt. While moving around this ring the bottles are filled, hopefully before they get back to the starting point. After they are filled the bottles are removed from the circular ring and placed on another linear conveyor belt that proceeds to the capping machine. Empty bottles cannot enter the filling machine faster than filled ones can be removed. Filled bottles cannot be removed from the machine faster than empty ones can enter. The problems and advantages of a circular bottling machine correspond in an analogous fashion to those of circular list structures in a computer.

EXERCISES

1. Write an algorithm to insert atoms into a multilinked list, such as the bank's customer accounts.

2. Write an algorithm to delete atoms from the multilinked bank customer account list.

3. How does the algorithm for insertion and deletion of atoms from a ring differ from the algorithm for insertion and deletion of atoms from a linked list?

4. Suppose that the customer's name requires 240 bits, the account number 30 bits, and the pointers 16 bits each. Compute the density of the multilinked list structure of the First National Bank pictured in Figure 3.17. What is the density for a list of 100 accounts? of 500 accounts?

5. *Backtracking* is the operation of scanning a list in reverse (finding successive backward pointers). Explain the difference between backtracking using a dense list, a singly linked list with forward pointers, and a doubly linked list.

6. Design a method for determining the account pointers of the Boulder Creek Bank. Refer to Figures 3.15 and 3.17.

4
RESTRICTED DATA STRUCTURES

4.1
Queues

The dense list and linked list structures discussed in Chapter 3 are general-purpose data structures. They can be used to store atoms or to maintain an order among atoms. However, they do require time and programming effort to update and scan. Certain applications may not need to access every atom on a list. Instead, they may want to retrieve only the first (or last) atom on the list. Any updates are made to the beginning (or end), and deletions are also made there. An example of this in a computer operating system is a list of jobs for processing. When the computer is available, it executes the first one on the list. Any new job is put at the end of the list. This type of list structure is a *queue*. It inserts atoms at the end of the list and deletes atoms from the beginning. A double-ended queue is a *deque* (pronounced "deck"), which makes possible insertions and deletions at both ends of the list. A deque is a more general structure than a queue, but it has limited usage in computer science. We will discuss it later.

A queue is a first-in, first-out (FIFO) structure needed for many applications of computer science. Fortunately, examples from everyday experience help us to understand queues. Every time you get in a line for service at a grocery store, bank, and so on, you are actually entering a queue. You gradually move toward the head of the queue as those in front of you finish their business and leave. Newcomers enter the line behind you, at the tail of the queue.

Most computer operating systems form queues at many places throughout the system. An example is an output queue. Normally, all the output produced in a single job resides on a disk until a printer is available. The entire collection of jobs waiting to be printed forms an output queue. With a nonpriority system and only one printer, the first job that finishes processing information is the first one printed. Depending on the speed of the printer and the amount of printing, the size of the queue grows or contracts. At times there may be no jobs on the queue. At other times there may be so many files to print that the Central Processing Unit (CPU) must delay processing new jobs so that it can complete some of the printing and decrease the size of the print queue.

How does the operating system implement the queue? There are many possible methods, several of which we will examine. One method is to reserve a block of memory that is big enough to hold the largest expected queue and to require the system to signal a queue overflow. For a printer queue the essential information that the queue must contain is a job identifier and a location of the job's output, although the queue may have other information,

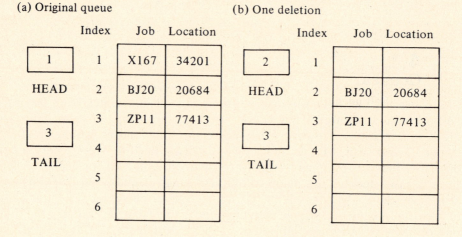

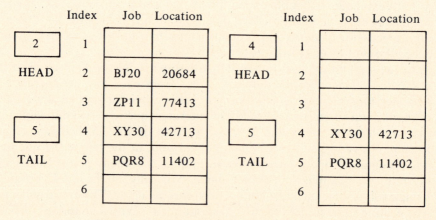

Figure 4.1 Output queue as a dense list

such as the number of lines of output. Initially the first jobs are placed at the beginning of the memory block. When the printer is free, the job removed from the queue is at the head of the list. Instead of moving all entries forward to close up the space left by the first one, we advance the HEAD pointer. The TAIL pointer shows the location of the last entry in the queue and any additions must follow it. Figure 4.1(a) shows the original queue; 4.1(b) illustrates the result of removing job X167 from the queue; and 4.1(c) shows the addition of two new jobs to the queue. In Figure 4.1(d) the first three entries of the memory block are vacant, but will not be used because the queue is gradually moving to the end of the area. If nothing is done, it will appear that the queue has overflowed when the reserved memory area is not nearly full. A solution to this problem is to move all the entries to the head of the list whenever an attempt is made to store something beyond the designated area. Figure 4.2 illustrates an example of this solution.

An alternative solution that eliminates moving the queue around is to make the queue memory area into a circular list. If an entry would cause an overflow, place it at the beginning of the area. Thus the end of the queue "comes before" the beginning if we examine the memory addresses involved. Logically, though, the end follows the beginning. (See Figure 4.3.) With this scheme we still need to worry about overflow, which would occur if the TAIL moved down and covered up the HEAD.

In either solution the update operations are similar. Basically they are given by the following algorithms. (The name QUEUE refers to the entire storage area of the queue; QUEUE[i] is the ith atom.)

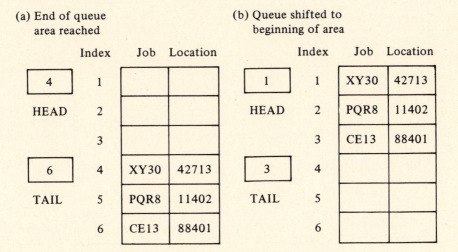

Figure 4.2 Moving queue to prevent overflow

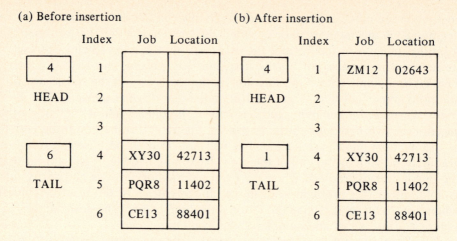

(a) Before insertion   (b) After insertion

Figure 4.3 Using a circular list to prevent overflow

Algorithm   INSERTQ

1. Comments: Basic algorithm to insert an atom on a queue. The queue is stored
    as a dense list. It assumes that there is room on the queue for the
    insertion and that the queue is not empty.

    Input: new entry   Data to be put on queue
    Output: nothing
    Variables: QUEUE   The queue
                TAIL   Index to tail (end) of queue

2. Increment TAIL index to point to next location: TAIL ← TAIL + 1

3. Insert new entry: QUEUE[TAIL] ← new entry

4. End INSERTQ

Algorithm   DELETEQ

1. Comments: Basic algorithm to delete an atom from a queue. It assumes that
    there is an atom to be deleted. Queue is stored as a dense list.
    Algorithm does not check HEAD to see if it is past the end of queue.

    Input: nothing
    Output: HDPTR   A pointer to the queue item that has been removed
    Variables: QUEUE   The queue
                HEAD   Index to beginning of queue

2. Remove QUEUE[HEAD]: set HDPTR ← pointer to QUEUE[HEAD]

3. Increment HEAD pointer: HEAD ← HEAD + 1

4. End DELETEQ

These queue INSERT and DELETE algorithms are not complete because they consider neither overflow nor *underflow* (trying to remove an atom from an empty structure). We therefore give the revised algorithms INSERT_QUEUE and DELETE_QUEUE. They assume that the queue is stored in a dense list and that the atoms are referenced by an index value. The largest index is N.

Algorithm INSERT_QUEUE

1. Comments: General algorithm to insert an entry onto a queue. The queue is stored as a dense list. Elements are moved to the beginning of the list if overflow occurs.

    Input: new entry   Data to be inserted
    Output: nothing   Unless an error is found
    Variables: HEAD   Index to beginning of queue; NIL if queue is empty
                  TAIL   Index to tail (end) of queue; NIL if queue is empty
                  N   Maximum length of queue

2. Do only one of the following cases:
    (a) If TAIL − HEAD + 1 = N, queue is full.
        No insertion made;
        Output 'Queue full'; EXIT INSERT_QUEUE
    (b) If HEAD is nil,
        Set HEAD ← 1; TAIL ← 0
    (c) If TAIL = N, end of dense list reached.
        Move QUEUE[HEAD] through QUEUE[TAIL] to QUEUE[1] through QUEUE[TAIL − HEAD + 1].
        Set TAIL ← N − HEAD + 1; HEAD ← 1

3. Increment TAIL: TAIL ← TAIL + 1

4. Insert new entry: QUEUE[TAIL] ← new entry

5. End INSERT_QUEUE

Algorithm DELETE_QUEUE

1. Comments: General algorithm to delete the head of a queue.
                  Queue is stored as a dense list.
    Input: nothing
    Output: HDPTR   A pointer to the queue item that has been removed.
    Variables: HEAD   Index to beginning of queue; set to NIL if queue is empty.
                  TAIL   Index to tail (end) of queue; set to NIL if queue is empty.

2. If HEAD is NIL, queue is empty.
    Output 'Queue Underflow'; EXIT DELETE_QUEUE

3. Remove QUEUE[HEAD]: set HDPTR ← pointer to QUEUE[HEAD]

4. Increment HEAD pointer: HEAD ← HEAD + 1

5. If HEAD > TAIL, then queue empty.
        Set HEAD ← NIL; TAIL ← NIL

6. End DELETE_QUEUE

The following Pascal program implements the INSERT_QUEUE and DELETE_QUEUE algorithms for a dense-list queue of job names. The main program marked with the comment "begin QUEUES program" asks the user to input a code signifying the operation to be performed along with any data required. It then calls either INSERT_QUEUE to insert a job, DELETE_ QUEUE to delete the first job in the queue, or PRINT_QUEUE to print the contents of the queue. The maximum size of the dense list was limited to ten to facilitate program debugging. Since the queue was implemented as a dense list, the HEAD and TAIL are not pointers representing machine addresses, but indexes into the dense list. The NIL pointer is represented by an index value of zero.

```
program QUEUES (INPUT, OUTPUT);          (* QUEUE Manipulation Algo-
                                            rithms                  *)

type
   CHAR1      = packed array [1 . . 1] of CHAR;
   CHAR4      = packed array [1 . . 4] of CHAR;
   MAXQ       = integer;
   Q_INDEX    = 0 . . 10;
var
   KODE        : CHAR;               (* Used in input to designate
                                        INSERT,   DELETE,   or
                                        PRINT operation           *)

   SKIP        : CHAR1;              (* Used in input for format
                                        control                   *)

   NEWJOB      : CHAR4;              (* Used in input; a job name to
                                        place on QUEUE            *)

   QUEUE       : array [1 . . 10] of CHAR4;
   N           : MAXQ;              (* Maximum QUEUE length;
                                        here = 10                 *)

   HEAD,                            (* Index to HEAD of QUEUE *)
   OLDHEAD,                         (* Index to previous head of
                                        QUEUE   that   had   been
                                        deleted                   *)

   NULL,                            (* Represents the NIL pointer *)
   TAIL        : Q_INDEX;           (* Index to TAIL of QUEUE    *)

procedure INSERT_QUEUE (NEWENTRY : CHAR4);
                                    (* Insert NEWENTRY into QUEUE *)
```

```
var
  I : Q_INDEX;
begin
  if TAIL − HEAD + 1 = N then              (* QUEUE full                    *)
    writeln ('QUEUE IS FULL; NO INSERTION MADE')
  else
    begin
      if HEAD = NULL then                  (* QUEUE empty                   *)
        begin
          HEAD := 1;
          TAIL := 0;
        end
      else
        if TAIL = N then                   (* QUEUE has reached end of
                                              dense list                   *)
          begin                            (* Move QUEUE to top of list *)
            for I := 1 to TAIL − HEAD + 1 do
              QUEUE[I] := QUEUE[HEAD+I−1];
            TAIL := N − HEAD + 1;
            HEAD := 1;
          end;
        TAIL := TAIL + 1;                  (* Update TAIL                   *)
        QUEUE[TAIL] := NEWENTRY;           (* Put new entry on QUEUE        *)
    end;
end;                                       (* End INSERT_QUEUE             *)


function DELETE_QUEUE : Q_INDEX;           (* Delete head of QUEUE         *)
                                           (*                              *)
                                           (* Output: index of previous
                                              head of QUEUE                *)
begin
  if HEAD <> NULL then                     (* QUEUE not empty              *)
    begin
      DELETE_QUEUE := HEAD;
      writeln (OUTPUT, QUEUE[HEAD], ' DELETED');
      if HEAD +1 > TAIL then               (* Deleted   last   item   on
                                              QUEUE                        *)
                                           (* QUEUE is now empty           *)
        begin
          HEAD := NULL;
          TAIL := NULL;
        end
      else
        HEAD := HEAD + 1;                  (* Advance  to  new  head  of
                                              QUEUE                        *)
    end
```

```
      else                                    (* QUEUE empty              *)
        begin
          DELETE_QUEUE := NULL;
          writeln (OUTPUT, 'QUEUE EMPTY');
        end
  end;                                        (* End DELETE_QUEUE         *)

  procedure PRINT_QUEUE;                      (* Prints QUEUE             *)
                                              (* Used in debugging        *)
  var
    I : Q_INDEX;
  begin
  if HEAD <> NULL then
    for I := HEAD to TAIL do
      writeln (OUTPUT, I, ' ', QUEUE[I])
  else
    writeln (OUTPUT, ' QUEUE EMPTY');
  end;                                        (* End PRINT_QUEUE          *)

  begin                                       (* Begin QUEUES program     *)
    NULL := 0;                                (* Initialize values        *)
    HEAD := NULL;
    TAIL := NULL;
    N := 10;
    writeln (OUTPUT,' ENTER "I" AND JOB NAME TO INSERT');
    writeln (OUTPUT,' OR "D" TO DELETE HEAD OF QUEUE');
    writeln (OUTPUT,' OR "P" TO PRINT QUEUE');
    writeln (OUTPUT,' OR EOF TO END');
    reset (INPUT, 'INTERACTIVE');             (* nonstandard Pascal        *)
    while not eof (INPUT) do
      begin
        readln (INPUT, KODE, SKIP, NEWJOB);
          if KODE in ['I', 'D', 'P'] then
            case KODE of
              'I':
                INSERT_QUEUE(NEWJOB);
              'D':
                OLDHEAD := DELETE_QUEUE;
              'P':
                PRINT_QUEUE;
          end                                 (* End of case              *)
        else
            writeln (OUTPUT, 'SPECIFY I, D, P, OR EOF');
    end;                                      (* End of while             *)
  end.                                        (* End QUEUES               *)
```

If we use a circular list for the queue and TAIL reaches the value N, we then want to reset TAIL to 1 during an insertion. The values of TAIL will thus be 1, 2, 3, . . . , N, 1, 2, . . . , and so on. The effect is similar to that of a circular linked list. However, our algorithms must perform the appropriate calculations to maintain a circular list, since there are no links to keep the list circular. To test for overflow, we must see if the new TAIL will be the same as the HEAD. Algorithms for implementing this type of queue follow.

Algorithm   INSERT_CIRC_Q

1. Comments: Algorithm inserts a new entry into a circular dense-list implementation of a queue.
    Input: new entry   Data to be inserted
    Output: nothing   Unless error detected
    Variables: HEAD   Index to beginning of queue; NIL if queue is empty
                TAIL   Index to tail (end) of queue; NIL if queue is empty
                N   Maximum length of queue

2. If TAIL − HEAD + 1 = 0 or N, queue is full.
    No insertion made.
    Output 'Queue Full'; EXIT INSERT_CIRC_Q

3. If HEAD is NIL, queue is empty.
    Set HEAD ← 1; TAIL ← 0

4. If TAIL = N, end of dense list reached.
    Set TAIL ← 0

5. Increment TAIL: TAIL ← TAIL + 1

6. Insert new entry: QUEUE[TAIL] ← new entry

7. End INSERT_CIRC_Q

Algorithm   DELETE_CIRC_Q

1. Comments: Delete the head of a queue. Queue is stored as a circular dense list.
    Input: nothing
    Output: HDPTR   A pointer to the queue item that has been removed
    Variables: HEAD   Index to beginning of queue; set to NIL if queue is empty
                TAIL   Index to tail (end) of queue; set to NIL if queue is empty
                N   Maximum length of queue

2. If HEAD is NIL, queue is empty.
    Output 'Queue Underflow'; EXIT DELETE_CIRC_Q

3. Remove QUEUE[HEAD]: set HDPTR ← pointer to QUEUE[HEAD]

4. If HEAD = TAIL, queue is empty.
    Set HEAD ← NIL; TAIL ← NIL;
    EXIT DELETE_CIRC_Q

5. Increment HEAD pointer: HEAD ← HEAD + 1

6. If HEAD > N, we have reached end of dense list.
     Set HEAD ← 1

7. End DELETE_CIRC_Q

In our discussion of the output queue we said that the queue might be empty sometimes and full other times. In the latter case, CPU operations would have to be delayed until some of the output could be printed. It is not desirable to delay the CPU just because the output queue may not be big enough. If we want to estimate the "best possible" size for the output queue, we must measure and estimate quite a few factors, such as the rate of insertion, rate of deletion, and service time. The study of queues is a well-established part of statistics, so we will not attempt to derive any of the following results.[1] If the average deletion rate μ is greater than the average insertion rate λ, the length of the queue will reach equilibrium. If the average insertion rate were greater than the average deletion rate, the queue would gradually grow longer and longer and would not stabilize. Assuming also that the service time is short for the majority of users and long for only a few, we get the average queue length l by the formula

$$l = \frac{\lambda}{\mu - \lambda}$$

Remember that this is only an average. The queue will sometimes be longer and sometimes shorter.

If we can, tolerate an overflow only z percent of the time, how much space must be set aside for the queue? The length k is given by the formula

$$k = \frac{\log z/100}{\log \rho} = \frac{\ln z/100}{\ln \rho}$$

and is derived by looking at the distribution of queue lengths. In the formula z is a percentage. Dividing z by 100 converts it to a decimal. The symbol ρ is the *traffic intensity* and is defined as λ/μ. In calculating k and l, very likely we will not obtain an integer answer. To be safe, we should therefore always convert the result to the next largest integer. For example, 1.3 should be converted to 2.

Example   A queue is expected to grow at the rate of 5 atoms/millisecond and to shrink at the rate of 8 atoms/millisecond. How much space should be needed for a queue that will overflow only 1% of the time?

---

[1] If you are interested, you may wish to take a course in queuing theory or operations research. For a derivation of queue length, see Leonard Kleinrock, *Queueing Systems, Volume 1: Theory* (New York: John Wiley & Sons, 1975).

The solution is obtained by substituting the numbers in the formulas. The average length l is

$$l = \frac{5}{8 - 5} = 1.667 \approx 2$$

However, the length so that overflow will occur only 1% of the time is

$$k = \frac{\log \dfrac{1}{100}}{\log \dfrac{5}{8}} \approx 10$$

which is quite a bit larger than the average length. These lengths assume a dense-list representation of a queue. If pointers are included, the amount of storage locations required would be greater. Suppose, for example, that the density is 1/3. The amount of space needed for a sparse list is $10 \div 1/3 = 30$ locations.

EXERCISES

1. How much memory is required to prevent overflow from ever happening to a queue?

2. Use Algorithms INSERT_QUEUE and DELETE_QUEUE and show (in diagrams) the items on the queue and the current values of HEAD and TAIL after each of the following updates. The code i is for an insert; d is for a delete. Assume that N = 6.

Job	Location	Code
AB11	43109	i
EF32	10741	i
MN01	82705	i
XR43	24722	i
ZR08	46210	i
AB11	43109	d
EF32	10741	d
BC39	21742	i
XT14	71348	i
LM75	24631	i

3. Using the following data, follow the instructions in Exercise 2.

Job	Location	Code
PR73	42765	d
XA41	21708	i
BR11	89034	i
XA41	21708	d
BR11	89034	d

4. Use the data in Exercise 2 but Algorithms INSERT_CIRC_Q and DELETE_CIRC_Q to update the queue.

5. Given that $\lambda = 4$ atoms/second and $\mu = 7$ atoms/second. Compute the average queue length l and the queue length k so that overflow will occur approximately 5% of the time.

6. Write subroutines to perform insertions and deletions from a queue. Design your own input format and output format and list the type of information that should go on the queue.

7. Design update algorithms for a queue that is stored in a linked-list structure. How are overflow and underflow handled?

8. Design update algorithms for a queue stored in a doubly linked list structure. How are overflow and underflow handled?

9. Give examples of other uses of queues, either in computer systems or in daily life.

10. Modify the Pascal program so that it operates on a queue stored as a circular dense list.

4.2
Stacks

Like the queue structure, the stack structure imposes an order on its elements. The stack's order is last-in, first-out (LIFO). The queue's order is first-in, first-out (FIFO). The place at which insertions and deletions are made determines the ordering. In a *stack,* insertions and deletions occur at the same end. called the *top* of the stack. In a queue, items are inserted at one end and then make their way through the queue to the other end, where they are deleted.

Although stacks are quite useful in computer science, we do find examples of them in everyday life. A railroad siding or a dead-end street impose an order on the cars that enter it. For example, in Figure 4.4 the first car in is

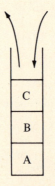

Figure 4.4 Diagram of a stack

A, the second B, and the third C. Since there is no other way out except the way the cars came in, the exit order is C, B, A—the reverse of the input order.

In computer operating systems, stacks are used for saving information so that it can be recovered later when it is needed. An execution-time stack is used during a program's execution to record pertinent information about the executing program (or function or subroutine). If the program calls a subroutine (or function), then the current execution environment is saved so that it can be restored when the called program exits. The information saved on the stack could include such things as name or address of the executing routine, address at which execution should resume, data values valid only for that execution state (for example, values of local variables), and contents of registers at the time execution is to be transferred to the other routine.

Thus, suppose that program MAIN calls INPUT and then CALC, and CALC calls GUESS and then VERIFY, as diagrammed in Figure 4.5. Our execution-time stack would go through the changes shown in Figure 4.6. The arrow points to the program or subroutine currently executing. Initially MAIN is the first program to execute. When MAIN calls INPUT [Figure 4.6(b)], INPUT is placed on the execution stack on top of MAIN. When INPUT ends execution, it is removed and MAIN resumes execution [Figure 4.6(c)]. Then MAIN calls CALC and CALC calls GUESS. Figure 4.6(e) shows the stack with GUESS, CALC, and MAIN. When GUESS terminates, CALC is at the top of the stack until it calls VERIFY [Figure 4.6(f)].

Placing or inserting an atom on a stack is also called a *push* operation, since it pushes all other atoms on the stack down by one position from the

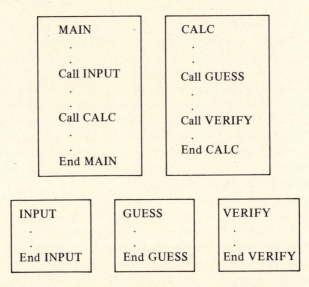

Figure 4.5 Diagram of main program and subprograms

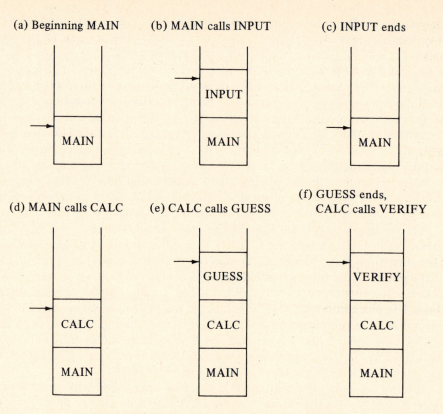

Figure 4.6 Execution stack of program MAIN

top. Removing or deleting an atom from the stack is also called a *pop* operation, since the top atom is popped off. With a particular implementation of a stack, there may not be any pushing, popping, or movement of atoms. Instead, the bottom of the stack may remain at the same location and a pointer or index may move to locate the top of the stack. Before we discuss the representation of stacks in storage, let us consider another application.

In a compiler, stacks assist in the scanning and analysis of statements. In particular, stacks facilitate the interpretation of arithmetic expressions so that a correct code will be generated. When a compiler scans an arithmetic expression, it must separate the operators from the variables and determine the order in which operations are to be performed.

To understand how this process works, let us study a simple expression involving only the $+$, $-$, $\times$, and $\div$ operators. There are no parentheses allowed and all variables (operands) are one character long. The rules for evaluation state that evaluation proceeds from left to right but that multiplication and division are done before addition and subtraction. Thus $A + B \times C$ will be interpreted as $A + (B \times C)$ and $X \div Y \times Z$ as $(X \div Y)Z$. It may be easy for us to follow the rules and evaluate expressions, but the computer's algorithm must work so that it has to scan only one character at a time. With this restriction in mind, we will rely on two stacks—one to save operands and one for operators that will not be needed until later. For example, in the expression A + B × C we must save the A+ until after B × C has been computed. Then we can take the result of the multiplication and add A.

Before stating the algorithm, we will go through the expression A + B × C − D and show how it is evaluated.

In the following discussion the arrow (↓) will show the character being scanned. The first two steps are to place A on the operand stack and plus (+) on the operator stack.

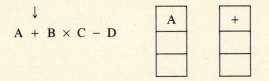

The next character is an operand, so it goes on the operand stack. Note that B goes on the top of the stack and A is now second on the stack.

$$\begin{array}{c} \downarrow \\ A \; + \; B \; \times \; C \; - \; D \end{array}$$

B		+
A		

The next character is an operator, so we must compare it with the top of the operator stack. A multiplication (×) operation is done before an addition (+) operation, so we say that it has higher precedence. An operator of higher precedence is placed on the stack.

$$\begin{array}{c} \downarrow \\ A \; + \; B \; \times \; C \; - \; D \end{array}$$

B		×
A		+

The C is an operand so it goes at the top of the operand stack, pushing B and A down.

$$\begin{array}{c} \downarrow \\ A \; + \; B \; \times \; C \; - \; D \end{array}$$

C		×
B		+
A		

The next character is minus ($-$), which has lower precedence than multiplication ($\times$) at the top of the operator stack. This is a signal that an operation can be performed. The operation involves taking the top operator ($\times$) with the top two operands (C and B) to evaluate B $\times$ C. We will call the result R_1. We now place the result on the top of the operand stack. Note that B, C, and the operator ($\times$) were removed from the stacks.

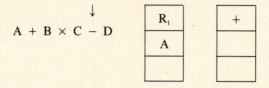

In addition, note that A and plus ($+$) have moved up. Since the operator stack has changed, we must compare minus ($-$) and the top operator ($+$). These operators have the same precedence, so we perform another operation, which is A $+$ R_1. Let us call the result R_2. As before, it is placed on the operand stack.

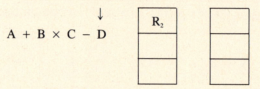

The operator stack is now empty, so we may place minus ($-$) on it and continue. The next character is D, which goes on the operand stack.

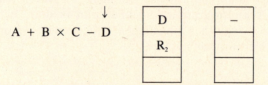

D is the last character in the expression, so we can evaluate all the remaining items on the stacks, always having one operator operating on the top two operands of the stack. In this case we compute R_3 as $R_2 - D$. Thus our evaluation has been

$$R_1 = B \times C$$

$$R_2 = A + R_1$$

$$R_3 = R_2 - D$$

The general algorithm for expression evaluation is Algorithm EXPR_ EVAL. Remember that it assumes that the expression has been correctly formed, that the expression does not begin with a plus ($+$) or minus ($-$) sign, and that there are no parentheses included. We are trying to illustrate the use of the stack, not the details of expression evaluation, which is a topic in a compiler design course.

Algorithm EXPR_EVAL

1. Comments: Evaluate an arithmetic expression using an operand stack and an operator stack. The algorithm assumes that the expression contains the operators $+$, $-$, $\times$, and $\div$, but no parentheses. All operands are one character long. The expression is assumed to be correctly formed and does not start with a leading $+$ or $-$ sign. The operators $\times$ and $\div$ have higher precedence than $+$ and $-$; the $\times$ and $\div$ have the same precedence; and the $+$ and $-$ have the same precedence.

   Input: an arithmetic expression

   Output: operations to evaluate the expression

2. Until the end of the expression is reached, get a character and perform only one of the cases (a) through (d):
   - (a) If the character is an operand, push it onto the operand stack.
   - (b) If the character is an operator and the operator stack is empty, place it on the operator stack.
   - (c) If the character is an operator and the operator stack is not empty, and the character's precedence is greater than the precedence of the top operator on the operator stack, place the character on the operator stack.
   - (d) If cases (a), (b), and (c) do not apply, then remove the top operator from the operator stack and the top two operands from the operand stack. Perform the operation s op f, where s is the second operand removed, f is the first operand removed, and op is the operator. Place the result back on the operand stack.

3. Repeat step 2(d) until both stacks are empty. If one stack empties before the other, an error has occurred because the expression was not correctly formed.

4. End EXPR_EVAL

Frequently we picture a stack as though it were built on top of a spring. Adding an atom to the top puts on more weight and pushes everything down. Removing an atom releases weight and the spring pushes everything up. The top atom is always at the same level, while the bottom of the stack may go up and down. In our implementation of a stack there will be no movement of the elements; instead we will use a pointer to locate the top. The vocabulary of stack operations, however, does reflect movement. The insertion of an atom to a stack is called *putting* or *pushing;* deletion of an atom is called *popping* or *pulling.*

In the implementation of a stack we may decide to reserve a block of memory so that the stack can grow and contract within that area. This is very similar to the implementation of a queue except that there is only a head pointer, which we will call TOP, instead of the HEAD and TAIL pointers of a queue. Also note one major difference: The TOP of the stack points to the last atom in the stack area. For better visual effect we might want to turn the stack upside down. Figure 4.7 illustrates the stack after insertion of A, B, and C, deletion of C, insertion of E and F, and finally removal of F, E, and B. Note that although B, E, and F appear in positions 2, 3, and 4, they are removed in the opposite order from the stack. First F will be popped, then E, then B. The stack is often called a *push-down stack* or *push-down store* because the insertion of an atom causes other atoms to go farther down on the stack.

The algorithms for putting an element on the stack (insertion) and popping an element from the stack (deletion) are given in PUSH and POP algorithms. They assume that the stack has room for N elements and that initially TOP is NIL.

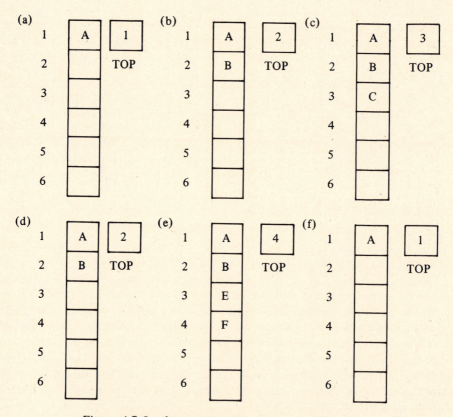

Figure 4.7 Implementation of a stack in a dense list

Algorithm   PUSH

1. Comments: Push or insert an atom onto a stack that is stored in a dense list. The maximum number of elements on the stack is N. An index locates atoms on the stack.

    Input: new atom   An atom to be pushed onto the stack
    Output: nothing   Unless an error occurs
    Variables: TOP   An index to top of stack; initialized at 0, representing an empty stack
                  STACK[I]   The Ith atom on the stack

2. Increment TOP: TOP ← TOP + 1

3. If TOP > N, then dense list is full.
    Output 'Stack Overflow'
    EXIT PUSH
    Else
    Put new atom on stack: STACK[TOP] ← new atom

4. End PUSH

Algorithm   POP

1. Comments: Pop or delete an atom from a stack that is stored as a dense list. An index locates atoms on the stack. The minimum index value is 1; zero denotes an empty stack.

    Input: nothing
    Output: top atom of stack
    Variables: TOP   An index to top of stack
                  STACK[I]   The Ith atom on the stack

2. If TOP is 0, there is nothing to delete.
    Output 'Stack Underflow'
    EXIT POP

3. Remove STACK[TOP]

4. Decrement top pointer: TOP ← TOP − 1

5. End POP

To prevent stack overflow, we can use a linked list, as pictured in Figure 4.8, to implement the stack structure and rely on a storage-management algorithm to supply us with additional space whenever an element is added. The advantage of this method is that our stack will occupy only the amount of space required at any one time and that it is expandable at execution time. When an atom is popped from the stack, we could return the unused space to the storage manager. We may thereby save space, but take more time to

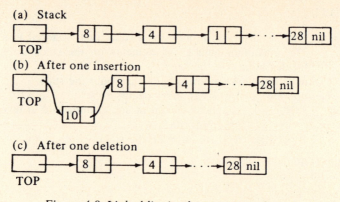

Figure 4.8 Linked-list implementation of stack

call the storage manager. Further applications of stacks appear in Sections 5.3, 7.2, 10.6, and 10.8.

EXERCISES

1. The PUSH and POP algorithms for a stack operate on a vector in which each entry in the stack occupies one position in the vector. Modify the algorithms so that they will update a stack that requires 10 elements of the vector for each entry in the stack.

2. Write PUSH and POP algorithms for a linked-list representation of a stack.

3. Trace through Algorithm EXPR_EVAL with the following expression:

$$X - A + B \div C \times Y \times P$$

4. Modify Algorithm EXPR_EVAL so that it will work with parenthesized expressions. If you pick the correct precedences for the left and right parentheses, the algorithm requires only minor changes.

5. In an operating system, which is better and why is it better: an output queue or an output stack?

6. Implement the PUSH and POP algorithms so that they operate on a stack of operators or operands one character long.

7. Should a stack be stored as a dense list or as a linked list? State the reasons for your decision.

8. Write a program to "evaluate" arithmetic expressions, as is done in Algorithm EXPR_EVAL. The output of the program could take the form of a series of *s op f*'s, where *s* and *f* are operands and *op* is an operator. This would show which operations are to be performed.

4.3
Deques

A deque is a generalization of a queue because it allows insertions and deletions at both ends. An alternative way to picture a deque is as two stacks joined together at the base. If there is a definite boundary between the stacks or if we can locate the last element of each stack, the deque provides a means for the stacks to share the same memory area. Each stack would have the potential to be as long as the maximum size allotted for the deque, yet we would need only one memory area for both stacks. Problems arise if both stacks are concurrently large, because they might overflow the area.

To implement a deque we will use a dense-list structure and will reserve a fixed amount of space. Since the deque can grow or shrink at both ends, we will need two pointers. Often we label the ends of a deque left and right or top and bottom, but to avoid any positioning of the deque, we will use END1 and END2. If the deque represents two stacks, END1 and END2 locate the tops of those stacks. To minimize the likelihood of the deque overflowing at either end, we should keep the atoms near the center of the list. Figure 4.9 illustrates a deque to which atoms are inserted and deleted. The numbers on the atoms refer to the order in which the atoms enter the deque. In step 4.9(e) there is no room to add any more atoms at END2. At this point, we can shift the deque toward the center of the area. Depending on the application, we may want to move the elements one position or enough positions so that there are an equal number of empty positions at each end of the deque. This movement is very similar to the implementation of the queue in which we moved the queue to the beginning of the area whenever we reached the end. As with queues and stacks, underflow can occur if we try to delete elements from an empty deque.

Because a deque allows insertions and deletions at both ends, we can use it as either a superqueue or superstack. In a queue, normally the first atom in is the first atom out. If we wanted to give a high priority to an atom, we could place it at the head instead of the tail of the queue. In a strict sense insertions cannot be made at the beginning of a queue, but they can be made in the beginning of a deque. In a stack the ordering is first one in, last one out. If the stack grows to a maximum size and overflow occurs, we could delete the atom at the bottom because it has been on the stack the longest. A deque in which output (deletions) can occur at only one end is called an *output-restricted deque*. A deque in which input (insertions) can occur at only one end is called an *input-restricted deque*.

An alternative way to represent a deque is to use either a singly or doubly linked list. Thus when we need to add an atom, we request additional storage from a routine that keeps track of available memory space. When atoms are deleted, we return the free space to the memory-management routine. To save

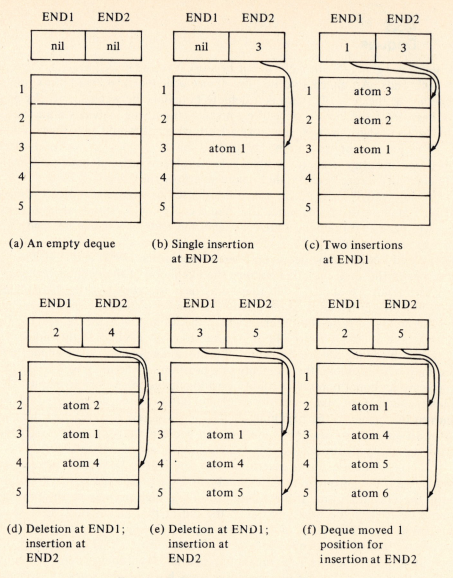

(a) An empty deque

(b) Single insertion at END2

(c) Two insertions at END1

(d) Deletion at END1; insertion at END2

(e) Deletion at END1; insertion at END2

(f) Deque moved 1 position for insertion at END2

Figure 4.9 Dense-list storage of deque

memory space we are using more time. Do not forget that links take added space.

Deques are the least restrictive structures discussed in this chapter; at present they are also the least useful in computer programming.

EXERCISES

1. Design an algorithm to insert an atom to END1 of a deque. Be sure to test for overflow. Will this algorithm work for END2?

2. Design an algorithm to delete an atom from END1 of a deque. Be sure to test for underflow. Will this algorithm work for END2?

3. If a deque occupies k spaces out of a total of n in the dense list, how many positions should there be at the top and bottom of the area so that the deque will be in the center of the list (approximately)?

4. Write algorithms to update a singly linked deque at either end.

5. Describe a doubly linked deque; write algorithms to update it at either end.

5
GRAPHS AND TREES

5.1
Definitions

In Chapters 1 through 4 we approached the study of data structures from the viewpoint of applications and implementation. Chapter 5 shifts emphasis momentarily to the study of structure as modeled by graphs. We must therefore introduce terminology and definitions that will describe data structures in a more general sense.

Once you understand the principles of graphs, the definitions of trees, files, and formal structures become easier to grasp. Moreover, theoretical terminology for graphs is commonly encountered in computer science, and you should become familiar with it.

Perhaps the most significant reason for studying graph theory as a topic in data structures is that it gives an abstract model of structure. An abstract model is useful in discussing data structures without concern for details of implementing them. In addition, we often discover similar structures in many applications. Therefore we will be less concerned for the moment about implementation and will continue with an abstract orientation provided by graph theory.

Graphs and trees are different from the linear structures we have already studied. We could list the atoms in a graph, but that would not tell us anything about the interconnections of the atoms. One of the best ways to represent a graph is with a diagram, because it pictures both the atoms and the relationships among the atoms.

A *nondirected graph,* such as the one shown in Figure 5.1, is a set of two types of objects: nodes and edges. A *node* is an element or atom of the graph; an *edge* joins two nodes. In the diagram in Figure 5.1 the nodes are the lettered circles and the edges are the lines. An edge has no direction; that is,

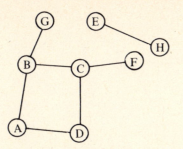

Figure 5.1 A nondirected graph

an edge between C and D is the same as the edge between D and C: It goes both ways.

If we name the nodes n_1, n_2, n_3, . . . , n_k, a *path* from n_i to n_j is a set of nodes n_i, n_{i+1}, . . . , n_{j-1}, n_j and edges such that there is an edge between successive pairs of nodes (n_i and n_{i+1}, n_{i+1} and n_{i+2}, . . . n_{j-1} and n_j). In Figure 5.1 there is a path from A to F. In fact, we could choose either path A–B–C–F or A–D–C–F. In this case we say that the graph contains an *alternate path,* since there is more than one path from one node to another. There is no path from H to F or from H to any other node except E. We say that this graph is disconnected. A *connected graph* has a path between any two nodes; a *disconnected graph* has at least two nodes with no path between them.

In Figure 5.1 there is a path that begins at node A, goes through nodes B, C, and D, and returns to node A. This path represents a *cycle:* a set of nodes n_i, n_{i+1}, . . . , n_j and edges for which there is an edge between successive pairs of nodes. All nodes are distinct except for the first and last, n_i and n_j, which are the same.

Applications of nondirected graphs include road maps, communications networks, and many other systems in which the components are connected by some means. Cycles in a map, network, or other type of graph indicate that there is more than one path from one place to another. Thus we can consider ways to find either the shortest path or the cheapest path. In the case of a road map, the cheapest path may be the fastest path, which is not always the shortest one. Another common problem is to devise a graph that will provide the "cheapest" network joining all the nodes. The cheapest graph should not contain any cycles, since a cycle represents a duplication of paths and part of the cycle could be erased. We will discuss this problem later.

We first need to consider *directed graphs,* or graphs in which the direction of the edge is important. An edge from A to B is *not* the same as an edge from B to A. Figure 5.2 represents a directed graph (*digraph*); the arrows show the direction of the edges. The definitions of cycle and path are the same for both digraphs and undirected graphs, but we must remember that an edge has direction. Thus in Figure 5.2 there is a path from E to D, but

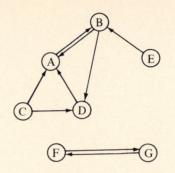

Figure 5.2 A directed graph

no path from D to E. The path B–D–A–B is a cycle, but there is no cycle containing the nodes D, C, and A. Ignoring directions, we could say that nodes D, C, and A form a loop or cycle, but considering direction, we find that all we have are two alternative paths from C to A. A digraph is *connected* if we ignore direction of the edges, and the resulting undirected graph is connected. Thus the graph in Figure 5.2 is not connected, but the subgraph containing nodes A, B, C, D, and E is connected. A *subgraph* contains a subset of nodes of the original graph together with a subset of the edges connecting those nodes.

In a directed graph we can talk about a successor node and the successor set of a node. A node n_i is the *successor* of node n_j if there is a path (not necessarily an edge) from n_j to n_i. Conversely, n_j is the *predecessor* of n_i. The *successor set* of a node contains all the nodes that are its successors. In Figure 5.2, D, B, and A form the successor set of E; C is the successor of no node since there is no edge pointing to it.

A tree is a special case of a graph. An *undirected tree* is an undirected graph that is connected and contains no cycles. A *directed tree* is a directed graph that has one node (called the *root*) whose successor set consists of all other nodes, and the graph contains no cycles and no alternative paths. An undirected tree has the special feature that any node can be the root; in a directed tree there is only one root. In Figure 5.3, (a) shows an undirected graph, (b) the same graph with C as the root, and (c) the same graph with E as the root. Contrary to nature, tree data structures are normally illustrated with the root at the top. Figure 5.4 illustrates a directed tree that has an arrow representing the direction of an edge. A directed tree can have only one root, even though the root may not appear at the top. In Figure 5.4, (a) and (b) represent the same tree.

A *subtree* is a subgraph of a tree with the property that the subgraph is itself a tree. In Figure 5.4 the subgraphs containing nodes A, E, and F, and nodes A, D, E, H, and I form subtrees. A subgraph is formed from nodes E, H, I, C, and G, but it is not a subtree because it is not connected.

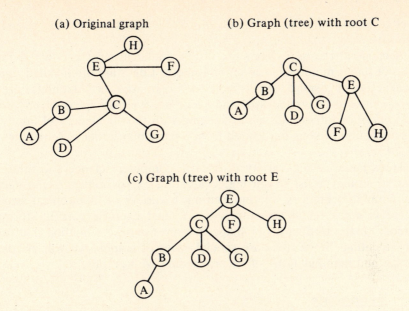

(a) Original graph   (b) Graph (tree) with root C

(c) Graph (tree) with root E

Figure 5.3 An undirected tree

In a directed tree we can count the number of edges pointing to a node (*indegree*) and the number of edges leaving a node (*outdegree*). The root has indegree of zero; the nodes with outdegree equal to zero are called *terminal nodes*. In Figure 5.4 the terminal nodes are D, H, I, F, and J. The level of a tree is a measure of the height or length of the longest path. The *level* of the root is 1. The level of every other node is one more than the level of its immediate predecessor. Thus nodes A and C in Figure 5.4 are at level 2; nodes H, I, and J are at level 4. When drawing trees we customarily place the root at the top and the nodes at the same level on the same horizontal line. (The level of the root is sometimes defined to be 0 instead of 1, but the

(a) Directed tree with root B   (b) Directed tree with root B not at top

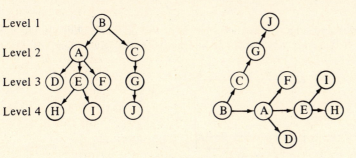

Level 1

Level 2

Level 3

Level 4

Figure 5.4 A directed tree

level of every other node is still one more than the level of its immediate predecessor.)

Because a tree as a data structure is similar in structure to a living tree, much of the terminology is common to both. Edges are often called *branches*. A collection of disjoint trees is called a *forest*. And terminal nodes are called *leaves*. Instead of predecessor and successor we may speak of *ancestor* and *descendant* or father and son in keeping with the vocabulary of family trees. The collection of sons descended from a particular father is sometimes called a *filial set*.

Trees have a wide variety of uses in computer science and other fields. Tree structures assist in sorting, evaluating expressions, making decisions, diagramming sentences, or representing a variety of hierarchical structures. A tree is not a linear structure, so we must exercise care in storing it in memory. If the tree varies in shape, we must choose a method that provides easy updating. Since trees are special types of graphs, we will discuss their storage and manipulation before proceeding with graphs.

EXERCISES

1. Describe an algorithm to convert nondirected graphs to directed graphs and an algorithm to convert directed graphs to nondirected graphs. If you start with an arbitrary graph and apply one algorithm and then the other, do you get back the original graph?

2. In the nondirected graph illustrated, list all the paths from B to E and all the cycles in the graph. Is the graph connected?

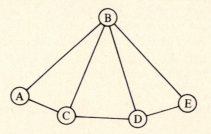

3. List the nodes and edges of each subgraph of the digraph below. Is the digraph connected? If there are any cycles or alternative paths, give examples.

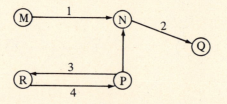

4. Redraw the following graph as a tree with root Y. List the nodes at each level.

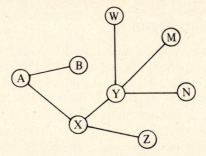

5. List all subtrees of the directed tree below. Give nodes and edges in each. How many nodes are in the largest subtree? How many nodes are in the smallest subtree?

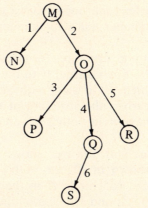

6. Give the indegree and outdegree of each node of the tree in Exercise 5.

7. Show how any forest can be converted into a single tree by the addition of a single node. How many edges need to be added? Can you do it without adding an extra node?

8. List the successor set of node B in the following digraph.

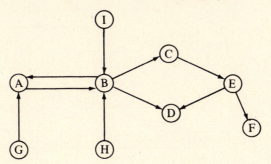

5.2
Storing and Representing Trees

We can store a tree by using a multilinked list and by including a pointer for each of the node's branches. The only difficulty with this method is that trees are frequently not regular. That is, not all the nodes have the same number of branches. For example, Figure 5.5 shows a tree and its representation as a multilinked list. Since the maximum outdegree (number of branches) is three, each atom has three pointers. Most of the nodes have outdegree less than three, so most of the pointers are null. We may be wasting space, but working with the tree is easier. The insert, delete, and update algorithms are less complicated than they would be if every atom had a different number of pointers. If efficient storage management were more important than ease of programming, we could let each node have a different number of links and include in each atom a number that tells the number of links. Terminal nodes would have no links.

One of the more common structures for trees is a form in which each node has at most two pointers. Such a tree, with outdegree at most two, is called a *binary tree*. The problem is to start with an arbitrary tree, store it as a binary tree, and then recover the structure of the original tree. We will see later that binary trees have many other advantages, so this problem is well worth solving.

Suppose that we start with the tree shown in Figure 5.6(a). One binary tree representation is shown in Figure 5.6(b), another is shown in Figure 5.6(c). Beginning with the root node A in the directed tree (a), we find its immediate successors B, C, and D. One of these, typically the leftmost, is put on the left branch of the binary representation. The right branch of A in the binary representation is reserved for a node at the same level as A in the

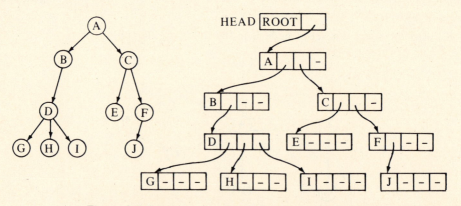

Figure 5.5 Storage of a directed tree as a linked list

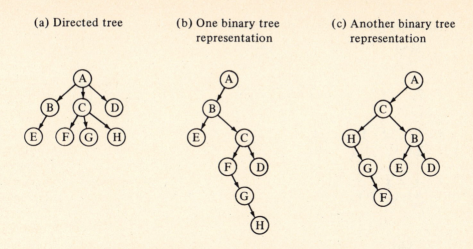

Figure 5.6 Representation of a directed tree as a binary tree

directed tree (there are none in this case). The same rules apply to each node. Consider, for example, node B in Figure 5.6(b). Its left branch points to node E and its right branch to C, because E is a successor of B in the directed tree and C is at the same level as B in the directed tree. Both B and C are immediate successors of A in the directed tree. Similarly, D is on the right branch of C in the binary tree and is at the same level as C and B in the directed tree. Nodes F, G, and H are all at the same level and are the immediate successors of C in the directed tree.

Figure 5.6(c) illustrates a binary tree created by placing C on the left branch of A instead of B. This shows that there are many binary tree representations of a given directed tree. To avoid confusion we will adhere to the method illustrated in Figure 5.6(b).

Consider the inverse problem: Given a binary tree, find the directed tree corresponding to it. First we locate the root node [L in Figure 5.7(a)] and make it the root node in the directed tree. The immediate successors of L in

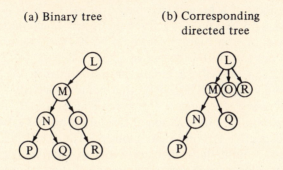

Figure 5.7 Binary tree and corresponding directed tree

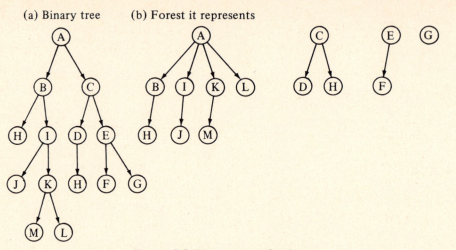

Figure 5.8 Binary tree of a forest

the directed tree are obtained from the left branch of L in the binary tree
(M) together with any nodes that can be reached from M by always following
the right branch in the binary tree (O and R). Since O and R have no left
successor in the binary tree, they have no successors in the directed tree.
The successors of M in the directed tree are determined in the same fashion.
They are N and Q, since N is on the left branch of M and Q is on the right
branch of N.

In Figure 5.7(a) the root L has only a left branch. What if the root had
a left branch and a right branch as shown in Figure 5.8(a)? Any node on a
right branch is at the same level as its immediate predecessor. Thus A, C,
E, and G are at the same level. Since A is the root node of the binary tree,
C, E, and G must also be root nodes. A tree has only one root; C, E, and
G are therefore roots for three distinct trees. The binary tree in Figure 5.8(a)
thus defines the forest shown in Figure 5.8(b). Even though the tree with root
G has only a single node, it is still a tree.

Since every node of a binary tree has at most two branches, we can use
a multilinked list with two links—one for the right branch and one for the left
branch. The binary tree in Figure 5.7 would be stored as shown in Figure 5.9.
The node name appears in the center of the atom.

In Pascal we can describe a binary tree structure in the following way.
First we define a data type NODE_PTR that points to TREE_NODE, the type
associated with every tree node. The type TREE_NODE is a record that
contains a name (NODE_NAME) ten characters long and LEFT and RIGHT
pointers to its respective left and right subtrees. The PARENT pointer locates
the node's parent, or immediate ancestor. None of the trees we have discussed
thus far had or needed parent pointers, but as we start manipulating trees we
will see where a parent pointer is useful.

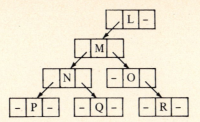

Figure 5.9 Binary tree of Figure 5.7 stored in multilinked list

```
type  NODE_PTR    =  ↑ TREE_NODE;
      CHAR10      =  packed array [1 . . 10] of CHAR;
      TREE_NODE   =  record
                        NODE_NAME   :  CHAR10;
                        PARENT      :  NODE_PTR;
                        LEFT        :  NODE_PTR;
                        RIGHT       :  NODE_PTR;
                     end;
```

After describing a tree's nodes, we may want to define a header to name the tree and locate the root of the tree. The variable TREEHEADER is a header for our tree. In addition to the tree's NAME and ROOT, the header contains LENGTH, the number of nodes in the tree. In our header we picked the maximum tree length as 500, but the choice was arbitrary.

```
var
   TREEHEADER   :  record
                      NAME     :  CHAR10;
                      ROOT     :  NODE_PTR;
                      LENGTH   :  0 . . 500;
                   end;
```

A tree, like a linked list, can impose an order on its elements. However, a tree can have a variety of orders imposed on its nodes depending on whether or not the left subtree is traversed before the right and whether or not all nodes at a single level are scanned before proceeding to the next. Since this is an important topic, we will delay it until a later section.

Putting nodes in the tree and obtaining storage for it depends on the application using the tree. A method for storing a binary tree in a dense list results from a particular numbering of the nodes. Figure 5.10 illustrates a binary tree with its nodes numbered in this manner. The root node is 1; its immediate successors are 2 and 3. In general, for any node i, its immediate successors are $2i$ and $2i + 1$. To store the tree in a dense list, we can use the list's indexes to locate nodes—their ancestors and descendants. For example, Figure 5.11 shows a tree and its storage in a dense list. The list names

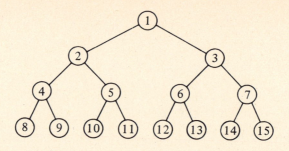

Figure 5.10 Binary tree numbered for dense-list storage

the nodes, but if each node had more data associated with it, then the list could contain pointers to each node's data. The tree in Figure 5.11 has five levels of nodes, so if the tree were as large as possible, it would have 31 nodes ($31 = 2^5 - 1$) instead of the 21 shown.

Traversing the tree (going from one node to the next) requires calculations to locate the next node in sequence. We can go up or down the tree, since we know that node i's parent has index $i/2$, discarding any remainder that may result from the division. These computations may seem trivial, but they can contribute a significant amount of traversal time if there are repeated multiplications or divisions. Also, if the dense list has many empty spaces, we may be wasting storage.

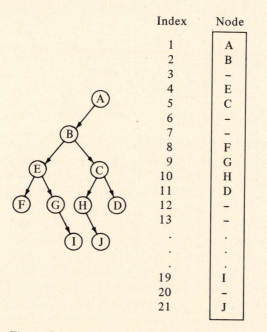

Index	Node
1	A
2	B
3	–
4	E
5	C
6	–
7	–
8	F
9	G
10	H
11	D
12	–
13	–
.	.
.	.
.	.
19	I
20	–
21	J

Figure 5.11 Binary tree stored in a dense list

For these reasons, we may want to use a multilinked list to store a binary tree as depicted in Figure 5.9. We usually think of the list's links as being pointers (addresses) to the location of other nodes. But if we are writing an application in FORTRAN or BASIC, then we can use a dense list to store the tree and use indexes instead of pointers.

Let us define three arrays (dense lists): NODE, LEFT, and RIGHT. We could use one two-dimensional array, but three arrays with names implying their use seems just as easy. Then we can place the node names in the array NODE and construct the LEFT and RIGHT indexes as the tree is built. For example, we can use the tree in Figure 5.11 to construct its indexed tree storage in Figure 5.12. We must also include an index to the root node, unless we know that it has an index of 1, as the binary tree in Figure 5.11 does.

Have we really saved any storage? If we assume that node names and left and right indexes each occupy one storage location, then Figure 5.12 uses 30

ROOT

```
┌─────┐
│     │
│  1  │
│     │
└─────┘
```

Index	NODE	LEFT	RIGHT
1	A	2	–
2	B	3	4
3	E	5	6
4	C	7	8
5	F	–	–
6	G	–	9
7	H	–	10
8	D	–	–
9	I	–	–
10	J	–	–

Figure 5.12 Binary tree using indexes for linked-list storage

locations, whereas Figure 5.11 uses 21. However, if Figure 5.11 had to allocate enough storage for a five-level binary tree, it would use 31 locations. In general, an n-level binary tree would require $2^n - 1$ locations by using the dense-list method of Figure 5.11. The multilinked-list method of Figure 5.12 uses $3m$ locations, where m is the number of nodes in the tree. To resolve the question of storage, we would need to compare $2^n - 1$ and $3m$. If the m nodes were arranged with one at each level, the maximum amount of storage required would be $2^m - 1$. The multilinked list is almost always better in terms of storage utilization, but certain applications, such as the heap sort in Section 7.2, make effective use of the dense-list representation of a tree.

We can probably understand storage implementations and tradeoffs better if we study an example. We will use a tree to order a set of numbers from smallest to largest. The tree is empty at first, but it will grow and change shape as the numbers get placed on the tree in numerical order.

Let us begin with a list of distinct numbers in a random order. Using the numbers, we construct a binary tree in the following fashion: The first number in the list becomes the root node of the tree. The next number in the list goes on the left branch if it is smaller than the root and on the right branch if it is larger. We take subsequent numbers in the list and compare them with the root node. If the number is smaller, it goes on the left branch. If it is larger, it goes on the right branch. If this process traces down a branch that has already been allocated to a previous number in the list, make the same comparison with the nodes at lower levels. That is, take the left branch if the number is smaller than the current node and the right branch if it is larger or equal.

Consider the set of numbers {325, 240, 572, 280, 108, 436, 720, 620}. We begin by making 325 the root node, as indicated in Figure 5.13(a). Since 240

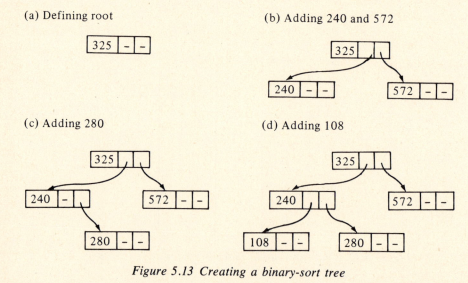

Figure 5.13 Creating a binary-sort tree

is smaller than 325, it goes on the left branch [Figure 5.13(b)]. And, of course, 572 goes on the right branch. How is 280 handled? Since 280 is smaller than 325, we proceed down the left branch to 240. We now compare 280 with 240 and put 280 on the right branch of 240.

Finish the list of numbers, making sure that the tree is equivalent to the one shown in Figure 5.14. Two representations of the tree are given—one in a linked-list diagram, the other in a dense-list representation of a linked list.

The purpose of this algorithm is to order (sort) a set of numbers. Does the set of numbers shown in Figure 5.14 appear ordered yet? To complete the algorithm, we must specify how the branches of the tree are traced to give the correct order to the set of numbers. This tracing process is referred to as *scanning* the tree, which will be discussed in Section 5.3. An alternative to the binary-sort tree is the B-tree, described in Section 10.10.

Before discussing tree operations, let us consider a general algorithm to build a tree. It can be adapted to work with trees stored in dense lists or in linked lists. It makes no assumptions about the order in which the nodes are to be placed. Those details can be added later to suit a particular application.

Algorithm BUILD_TREE

1. Comments: General algorithm for building a binary tree. No details about the tree's search order or storage are assumed.
   Input: new data To be added to tree
   Output: nothing

(a)   (b)

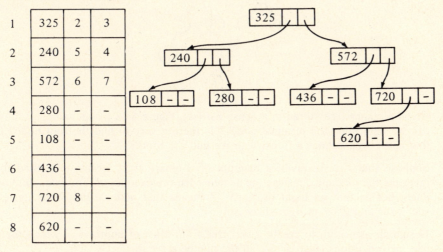

Index Value Left Right

Index	Value	Left	Right
1	325	2	3
2	240	5	4
3	572	6	7
4	280	–	–
5	108	–	–
6	436	–	–
7	720	8	–
8	620	–	–

Figure 5.14 Completed binary-sort tree

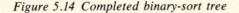

2.   Obtain or allocate storage for new node.

3.   Search tree and determine place where new data should go.

4.   Place the new node's data in the tree storage area obtained in step 2.

5.   End BUILD_TREE

EXERCISES

1.   Convert the following directed tree into a binary tree. How many levels does the binary tree have? the directed tree?

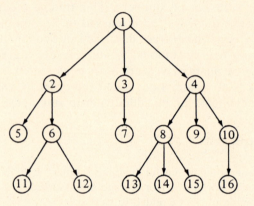

2.   Convert the following binary tree into a directed tree or forest.

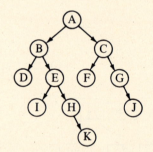

3.   Given a binary tree stored in a dense list (see Figure 5.12 for an example), how can we determine the terminal nodes? Is there a way to find the root node? If the binary tree defines a forest, can we find the roots?

4.   Suppose that we are given a diagram of a binary tree. Design a method for inputting the tree and constructing its dense list representation. For example, you might decide that an input of X, W, Z means that node X has W as its left successor and Z as its right successor.

5.   Design an algorithm to create a sort tree from a list of numbers in which there may be repetitions (that is, not all the numbers are distinct).

6. Design an algorithm to print the nodes of a sort tree, such as the tree shown in Figure 5.14, in descending order (from largest to smallest).

7. Expand the BUILD_TREE algorithm described in the text to create a binary-sort tree.

8. Implement the algorithm in Exercise 7.

9. Show how the binary-sort tree in Figure 5.14 would be stored in a dense list without any pointers. That is, the root has index 1, its successors have indexes 2 and 3, and so forth. Use Figure 5.12 as an example.

10. Suppose that a binary tree has 25 nodes. What is the smallest number of levels it can contain? What is the largest number of levels? In general, if a binary tree has n nodes, what is its smallest number of levels? its largest number?

11. Write an algorithm to convert a binary tree into an equivalent directed tree.

12. Design an algorithm to convert a directed tree into a binary tree using the method represented by Figure 5.6(b).

5.3
Operations on Trees

Scanning a tree is necessary when searching for a particular node on the tree or determining where an insertion should occur. A tree scan is more complicated than a list scan because a tree is not a linear structure. One of the easiest ways to search a list is to begin with the first element and continue from first to second on through the list to the last element. The branches on a tree make a tree search more difficult. At every node in a binary tree we usually have the choice of taking either the left branch or the right branch. In the tree scan we must be methodical and yet trace all the branches without visiting any node twice and without omitting any node.

The fundamental part of a binary tree structure is a node with two branches (left and right). In places where one or both branches may not point to another node, the branches are not usually drawn. Considering this fundamental part of the structure, we see that there are three places to begin the search: at the root node (N), the left branch (L), or the right branch (R). Listing all possible combinations of L, N, and R, we find that there are six different scanning orders: LNR, LRN, NLR, NRL, RNL, and RLN. L and R point to subtrees that have the general form shown below and that must also be scanned. In the following discussion we will select three of the six possible scan orders and apply them recursively to each node of the tree, usually beginning with the root pointed to by the ROOT pointer.

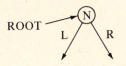

An NLR scan of a binary tree considers first the root node, then the left subtree, and finally the right subtree. In each subtree, the scan follows the same NLR rule: node, left subtree, right subtree. The NLR scan is also known as a *preorder* scan. The name comes from the fact that the root is scanned first, or before its subtrees' scan. To understand the implementation of this rule, consider the sample tree in Figure 5.15. Begin with node A and then work with the left subtree, which consists of nodes B, D, E, and H. Looking at only the left subtree, we scan its root node B, left subtree (node D), and right subtree (nodes E and H). Since D is a terminal node, we have completed scanning the left subtree of B and can now scan the right subtree. In NLR order we first examine E (root node of right subtree of B) and then H. So far, we have considered nodes A, B, D, E, and H in the NLR-order scan and proceed to the right subtree. Processing it with the same NLR scan rule produces the ordering C, F, G, I, and J. Thus the complete scan is A, B, D, E, H, C, F, G, I, J.

An LNR scan considers the left subtree, the node, and then the right subtree. *Inorder* is another term commonly used for an LNR scan. The name comes from the fact that the root is scanned in between the scan of its left and right subtrees. For the tree in Figure 5.15, the LNR scan produces the ordering D, B, H, E, A, F, C, I, G, J. The LRN scan considers first the left subtree, then the right subtree, and finally the node. With the same tree (Figure 5.15), the LRN scan yields D, H, E, B, F, I, J, G, C, A. The LRN scan is also known as a *postorder* scan, since the root is scanned after its left and right subtrees.

The LRN order is frequently used by a compiler as a way of internally representing the source program. The compiler scans the source program, recognizes the instructions and their components, generates in LRN order an intermediate text form of the instructions, and then converts the intermediate text into object code. The LRN order is a way of describing the components of a source statement or expression so that their meaning is not ambiguous.

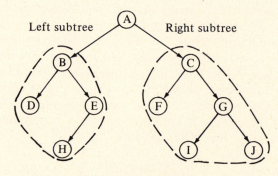

Figure 5.15 A binary tree

For example, the arithmetic expression

$$c - t * 1.06$$

can be diagrammed with the following tree, which says that t and 1.06 are to be multiplied and that the result is to be subtracted from c.

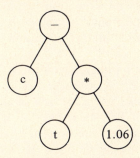

Applying an LRN order to the tree, we convert it into a linear list:

$$c\ t\ 1.06 * -$$

This list can then be scanned and the correct object code can be produced by the compiler. The linear form together with the notion of LRN order explicitly describes the expression.

An alternative to LRN order is NLR order, which for our example would place the operators before the operands. The NLR order for our expression is:

$$-c * t\ 1.06$$

This NLR, or preorder, form is also known as *Polish notation*, since it was introduced by the Polish logician Łukasiewicz.

Which of the scans would we use to print the nodes of the sort tree in Figure 5.14 in ascending order? Recalling the way that the tree was constructed, a value less than the root went on the left branch, and a value larger than the root went on the right. In the sort tree all nodes on the left subtree are thus smaller than the root and all nodes on the right subtree are larger than the root. Using an LNR scan we can construct the list of numbers, ordering from smallest to largest.

Since trees have pointers from nodes to their successors, it is easy to travel "down" the tree from node to node until we reach a terminal node. The question then becomes: How do we get back "up" the tree to go down another branch? We can look at the diagram of a tree and easily determine a root's predecessor, as well as its successors. If the tree nodes have parent pointers, locating predecessor nodes is then facilitated. Another alternative is to use a stack and store the path that goes down the tree.

The following LNR_SCAN algorithm does not appear to use a stack, but it is *recursive*. That is, it calls itself. Programming languages that allow recursive programs use a stack to save the current execution state before the program calls itself. Then, as it returns or exits, the previous execution state is retrieved from the stack and execution resumes. If a program calls itself 10 times, there will be 10 different execution states on the stack. One by one the program would exit each of the execution levels before it terminated completely.

The LNR_SCAN algorithm begins with the ROOT node as input and then starts examining the left subtree of the ROOT. If the root has a left subtree, then LNR_SCAN calls itself, using the top of the root's left subtree as the "root" under consideration. The algorithm thus works on smaller and smaller left subtrees until it has reached the bottom of the tree. It then outputs the left subtree node currently being scanned. It next outputs the left subtree's root node and begins work on the right subtree. The right subtree scan is similar to the left subtree scan. First the left side is scanned, next the root, and then the right.

Algorithm   LNR_SCAN

1. Comments: LNR recursive scan of a binary tree. To scan an entire tree, it must be given the tree's root node.
   Input: P   Pointer to node being scanned
   Output: the tree's nodes   In LNR order
   Variables: NEXTP   Pointer to next node to be scanned
                 LEFT   A node's pointer to its left subtree
                 RIGHT   A node's pointer to its right subtree

2. If P is nil, then
      Output 'Error: LNR_SCAN Called with Nil Pointer'
      EXIT LNR_SCAN

3. Scan left subtree:
      NEXTP ← LEFT pointer of P
      If NEXTP is nil, then
         Output node P, the node pointed to by P
      Else
         Call LNR_SCAN recursively using NEXTP for input
         Output node P upon returning

4. Scan right subtree:
      NEXTP ← RIGHT pointer of P
      If NEXTP is not nil, then
         Call LNR_SCAN recursively, using NEXTP for input

5. End LNR_SCAN

To understand the algorithm, we will scan the tree in Figure 5.15. Initially P points to A, the root node. Since P is not nil, we scan A's left subtree and set NEXTP pointing to B. NEXTP is not nil, so we call LNR_SCAN with the pointer to B. At this point we draw one solid line in Figure 5.16 (our trace record). The line denotes that we have switched from one execution state to another. The value of P (our input to the algorithm) is a pointer to B. Before calling LNR_SCAN, we note that we were in step 3. Then when we exit from LNR_SCAN, we can resume execution where we left off. We proceed as before, this time calling LNR_SCAN with the pointer to D. During our third call, we reach the terminal node D and output it. Then we exit LNR_SCAN and return to our second level of execution, with P pointing to B and NEXTP pointing to D. Here we output B and begin to scan the right subtree. You should continue tracing the algorithm in this manner to verify that the LNR order of D B H E A F C I G J is achieved. It is possible to write an LNR-scan algorithm that is not recursive, but we leave that as an exercise for you to do.

The LNR order that we created in the example in Section 5.2 in which we sorted a set of numbers gave us the ordering L < N ≤ R. This means that all values on the left subtree are less than the root node, and the root node is less than or equal to all values on the right subtree. Values equal to the root can go on either the left or the right subtree, but we should pick one side

---

CALL 1

    P points to A
    NEXTP points to B
    Step 3 — CALL LNR_SCAN (pointer to B)

CALL 2

    P points to B
    NEXTP points to D
    Step 3 — CALL LNR_SCAN (pointer to D)

CALL 3

    P points to D
    NEXTP is nil
    Output 'D'
EXIT 3   End (exit) LNR_SCAN

CALL 2

    Output 'B'
    NEXTP points to E
    Step 4 — CALL LNR_SCAN (pointer to E)

Figure 5.16 Partial trace of Algorithm LNR_SCAN on Figure 5.15

and use it consistently. The way our tree was constructed, the first value became the root node. If the first value were not close to the middle of all values, then our tree might not have a balanced or very even shape. If we were so unlucky as to pick the smallest value first and make that the root, then our LNR tree might look like a linked list with pointers going off to the right. If our tree is going to grow and change shape dynamically, then we may want to consider a different ordering on the nodes. Unfortunately, no matter which ordering we pick, there will be some set of data for which the tree is not balanced.

Using a recursive algorithm avoids the problem of tracing back up the tree. Similarly, a nonrecursive algorithm that uses a stack to record nodes as one goes down the levels of the tree can backtrack by popping nodes off the stack.

An alternative way to backtrack during a tree scan is to construct the tree with additional pointers. These pointers are called *threads* because they tie the nodes together, thereby assisting in the scan. The threads occupy places that would normally be nil pointers. The size of a threaded tree is the same as that of a nonthreaded tree, and scanning requires no stack. In general, for whatever order is imposed on a tree, the threads reflect that order. The left thread of a node points to the node's predecessor in the chosen order. A node's right thread points to its immediate successor.

Figure 5.17 illustrates the tree in Figure 5.15 with LNR threads included. Dashed lines represent the threads. The right thread of a node points to the node's successor in the LNR scan; the left thread points to the node's immediate predecessor. Algorithm SCAN_THREAD scans a tree with LNR threads. It begins with a pointer to the root node. If the tree has a left branch, we begin by going down the left until we reach the leftmost node, which begins the LNR scan. Then we begin the basic loop, which results in a complete scan of the tree. First we go down a right branch. If that branch is a thread, it signifies that the node it points to is to be scanned (output) next. If the right branch is not a thread, we go off following the left branches

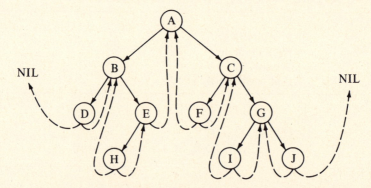

Figure 5.17 A binary tree with LNR order threads

until we reach a node whose left branch is a thread. We output it and repeat our loop.

Algorithm   SCAN_THREAD

1. Comments: Scan a binary tree threaded in LNR order. The scan begins at the root node.
    Input: P   Pointer to root node (current node under consideration)
    Output: nodes   In LNR order
    Variables: LEFT   Pointer to left branch of node (if negative, this denotes a thread)
                 RIGHT   Pointer to right branch of node P (if negative, this denotes a thread)

2. Search down left of tree for first node in LNR order:
        If LEFT of P is not a thread,
            Repeat:
                P ← LEFT of P
            Until LEFT of P is a thread

3. Output node P

4. If RIGHT of P is not nil,
    Repeat the following until RIGHT of P is nil:
    (a) Go down right branch or thread:
             P ← RIGHT of P
    (b) If we traversed a thread (Is P a thread?),
             Output node P
         Else
             Go down left branch until LEFT of P is a thread:
                 Repeat:
                     P ← LEFT of P
                 Until LEFT of P is a thread
             Output node P

5. End SCAN_THREAD

In Figure 5.17 two of the nodes have nil threads. Node D is the first node in LNR order; node J is the last. An implementation of a tree with threads might use a negative value to denote a thread. In essence, we use a positive sign to denote a pointer and a negative sign for a thread. Figure 5.18 shows the tree from Figure 5.17 stored in linked-list form, using indexes instead of pointers. Here a negative index represents a thread. Alternatively, if it were easy to do and did not require much additional storage, every left and right branch could have a flag to denote whether the branch was a pointer or a thread.

The insertion of a node into a threaded tree is fairly easy and straightforward. Algorithm INSERT_THREAD tells how to insert a node into an LNR-threaded tree. It assumes that we have already located where the new node

ROOT

```
┌─────┐
│     │
│  1  │
│     │
└─────┘
```

Index	NODE	LEFT	RIGHT
1	A	2	3
2	B	4	5
3	C	6	7
4	D	–	–2
5	E	8	–1
6	F	–1	–3
7	G	9	10
8	H	–2	–5
9	I	–3	–7
10	J	–7	–

Figure 5.18 A threaded tree stored using indexes for links

should go and that it will go on the right branch of a particular node. The new node is threaded back to P, its predecessor. And, if necessary, a thread is adjusted to point to the new node.

Algorithm   INSERT_THREAD

1. Comments: This algorithm inserts a new node on an LNR-threaded tree. It assumes that the new node is to go on the right branch of a particular node P. A similar algorithm could insert the node on the left branch.

    Input: new node   A new node to be inserted in tree
    P   Pointer to node on whose right subtree the new node is to go

    Output: updated tree
    Variables: LEFT   Pointer to a node's left subtree
    RIGHT   Pointer to a node's right subtree
    N   Pointer to storage for new node

2. Get storage for new node:
    N ← pointer to storage for new node

3. Put new node on P's right branch:
    RIGHT of N ← RIGHT of P
    RIGHT of P ← N

4. Left thread of new node points to P, its LNR predecessor:
    LEFT of N ← thread to P

5. If RIGHT of N is a pointer, change left predecessor thread from P to N:
    Find N's successor in LNR order.
    Set LEFT of N's successor ← N

6. End INSERT_THREAD

So far we have discussed inserting nodes on a tree and scanning the tree. Searching the tree for a particular node is similar to scanning, since we can use the tree's order to assist us in deciding which branch to take. One topic we have neglected (perhaps because it is a little more difficult) is that of deleting a node from a tree. If the node is a terminal node, we must locate its parent and make the appropriate pointer nil. In this way, the node being removed is no longer pointed to.

To delete a nonterminal node, we must decide where to place its left and right subtrees. In a tree whose nodes are placed in a particular order, we must be careful to attach the deleted node's subtrees in the correct place. This may require shifting a subtree up to occupy the place where the deleted node was. To illustrate deleting a node, we will develop an algorithm to delete the first node from an LNR-ordered tree. If the tree contained events to schedule at a particular time or within a particular priority scheme, then this algorithm would locate the event to schedule next.

Algorithm   DELETE_FIRST

1. Comments: Delete the first node from an LNR-ordered binary tree.
    Input: nothing
    Output: pointer to first node
    Variables: P   Pointer
                  LEFT   Pointer to node's left subtree
                  RIGHT   Pointer to node's right subtree
                  PARENT   Pointer to node's parent
                  NEWLEFT   Pointer to new first node; this node will be deleted
                                 next
                  ROOT   Pointer to root node

2. Point to root of tree:
    P ← ROOT

3. If root is nil, then

Output 'Tree is empty'
EXIT DELETE_FIRST

4. Search for bottom left atom:
    Repeat while LEFT of P is not nil:
        P ← LEFT of P

5. Save P's LEFT, RIGHT and PARENT pointers:
        P_LEFT ← LEFT
        P_RIGHT ← RIGHT
        P_PARENT ← PARENT

6. Decide what the LEFT pointer of P's parent should be:
        If P_RIGHT is nil, then P has no right subtree,
            Set NEWLEFT ← nil
        Else
            Shift up P's right subtree:
                NEWLEFT ← P_RIGHT
            Set the parent of P's right subtree to P's parent:
                PARENT of NEWLEFT ← P_PARENT

7. If P is not the ROOT (P_PARENT is not nil),
        Set LEFT of P's parent to its new left value:
            LEFT of P_PARENT ← NEWLEFT
    Else ROOT was deleted
        Set ROOT to new left:
            ROOT ← NEWLEFT

8. End DELETE_FIRST

To illustrate the effects of the algorithm, let us delete the first node from the tree in Figure 5.19. Its original shape is shown in Figure 5.19(a). Node 2 is the first node to delete. Since it is a terminal node, we merely remove it from the tree; the result appears in Figure 5.19(b). Removing the first element from Figure 5.19(b) results in Figure 5.19(c), where node 5's right subtree is

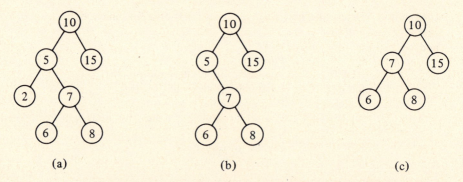

(a)   (b)   (c)

Figure 5.19 Deleting the first node from an LNR tree

shifted up to replace node 5. Deleting a node other than the first node is left as an exercise for you to do.

We have seen that the tree structure, together with its representations and algorithms, has a wide variety of applications in computer science. These applications range from sort to compile, from simulation of event lists to file indexes. Throughout the text we will continue to encounter trees, but first we study a generalization of a tree—the graph.

EXERCISES

1. List the nodes of the following tree in NLR, LNR, and LRN order.

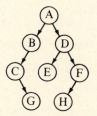

2. Write an algorithm to scan a tree using the NLR rule.

3. Convert the following tree into a binary tree. Then list the nodes in NLR order and LNR order. What is the difference between these two orderings applied to the original tree?

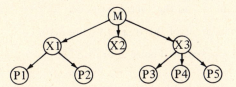

4. Write an algorithm to do an LRN scan of a binary tree.

5. Can the rules of NLR, LNR, and LRN order scan be modified for a tree each of whose nodes has at most three branches? If so, choose one of the orderings and write the appropriate rule.

6. Draw the LNR-order threads for the tree in Figure 5.15. How does your diagram differ from the one in Figure 5.17? With which node does the scan begin? How can we locate it easily?

7. Write an algorithm to scan a threaded tree with NLR-order threads.

8. If a binary tree has n nodes and is stored as a dense list, how many left and right pointers will there be? Of the left and right pointers, how many will not be nil?

9. What major problem occurs in tree scanning that is partially overcome by threading the tree? How do threads change the definition of a tree?

10.   Design an algorithm to create a sort tree inserting LNR-order threads as it is created.

11.   Write an algorithm to input nodes and create a tree with nodes in NLR order.

12.   Write a program to implement the algorithm in Exercise 11. Include a print routine to print the tree.

13.   Write an algorithm to input nodes and create a tree with nodes in LNR order.

14.   Insert node N's key value after node C in the tree in Figure 5.17.

15.   Insert node X's key value after node E in the tree in Figure 5.17.

16.   Modify Algorithm INSERT_THREAD so that it places the new node on the left branch of P.

17.   Draw a tree and then scan it by tracing through Algorithm SCAN_THREAD.

18.   Write an algorithm to scan a tree threaded with NLR-order threads.

19.   Draw the tree in Figure 5.17 using NLR threads instead of LNR threads.

20.   Write an algorithm to delete any node from an LNR-ordered tree.

21.   Write an algorithm to delete the first node from an NLR-ordered tree.

22.   Write an algorithm to delete any node from an LRN-ordered tree.

23.   Draw trees as examples to show what happens when a particular node is inserted in the tree. Do this for: (a) an NLR-ordered tree, (b) an LNR-ordered tree, (c) an LRN-ordered tree.

24.   Design an algorithm to find the successor node of a tree threaded in LNR order.

25.   Give the LNR scan order of the tree in Figure 5.10.

26.   List the nodes of the tree in Figure 5.10 in NLR order and in LRN order.

5.4
Manipulating Graphs

Since we have already discussed basic concepts of graphs, we can turn our attention to some applications. The study of graphs is a complete topic in itself, so we will mention only a few applications in this book.

A common example of a graph is a road map; the nodes are cities or intersections and the edges are roads or streets. A graph (or digraph) in which the nodes have weights is a *weighted graph* (or *weighted digraph*). On a road map the weights might be speed limits, average traveling times, or lengths of roads. On most graphs we label the nodes for easy identification. If nodes have more than one edge joining them, it is convenient to label the edges as well.

Recall that a path is a collection of nodes a_i, a_{i+1}, . . . , a_j such that there is an edge between successive pairs of nodes. In a *simple path* all nodes are unique. The shortest path between two points is necessarily a simple path. For example, in Figure 5.20 path a_3, a_4, a_5, a_3, a_6, a_7 contains a cycle and therefore is not a simple path. However, it can be reduced to the simple path a_3, a_6, a_7 by removing the cycle. Some of the oldest problems in graph theory are concerned with finding a simple path containing every node of the graph. Such a path is called a *Hamiltonian path*. It is named after Sir William Rowan Hamilton, who first introduced the problem in the form of a puzzle (see Exercise 1 at the end of this section). In a more practical form this puzzle occurs frequently enough so that considerable computer time is spent searching for solutions. A recent application involves a computer search to determine a Hamiltonian path that passes through every state capital of the United States and minimizes the total amount of travel. Although computers are ideally suited to this type of search, the amount of computation involved can be enormous, even with a modest number of nodes. Because the number of edges can increase rapidly, it is worthwhile to study storage of graphs before investigating applications.

A graph is completely determined by a list of pairs of nodes, where a pair indicates an edge between the nodes. Thus (A, B) indicates an edge between nodes A and B. The order is not important for a nondirected graph, so (A, B) is the same as (B, A). To specify a digraph, use pairs of nodes but add the rule that (X, Y) means an edge from X to Y, but not necessarily an edge from Y to X.

One straightforward way to store a graph is by storing a list of the nodes and a list of the node pairs that determine the edges. Since this method requires a lot of memory, we will use conventions to save space and to facilitate work with the graph. Instead of listing node pairs as (A, B), (A, D), (A, F), we will shorten them to (A: B, D, F). This notation says that there are three edges which begin at A and end at B, D, and F, respectively.

Figure 5.21 illustrates a graph and its tabular representation using this notation. The table says that there are edges from the first entry in a row to

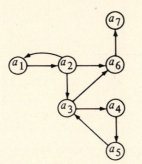

Figure 5.20 A digraph

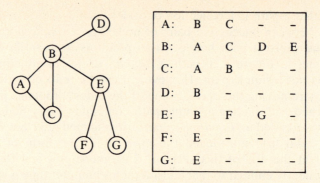

A:	B	C	–	–
B:	A	C	D	E
C:	A	B	–	–
D:	B	–	–	–
E:	B	F	G	–
F:	E	–	–	–
G:	E	–	–	–

Figure 5.21 A nondirected graph and tabular representation

each of the other entries in the row. The introduction of pointers allows us to condense the table into a list structure. In Figure 5.22 we see a possible implementation. All the structure of the original graph is contained in these lists, but it is becoming more obscure. The INITIAL and TERMINAL tables list the initial and terminal nodes of each edge in the graph. For example, rows B,3 and C,7 of the INITIAL table tell us that there are four (7 − 3 = 4) edges that begin at B and end at nodes A, C, D, and E. The pointer associated with B (a 3) locates the first of these four nodes in consecutive locations in the TERMINAL table. The last entry in the INITIAL table serves the purpose of determining the number of edges that begin at G. Note that Figures 5.21 and 5.22 list every edge twice. It is possible to eliminate this redundancy, but manipulation of the graph becomes more difficult.

Node	Pointer
A	1
B	3
C	7
D	9
E	10
F	13
G	14
–	15

INITIAL

Index	
1	B
2	C
3	A
4	C
5	D
6	E
7	A
8	B
9	B
10	B
11	F
12	G
13	E
14	E

TERMINAL

Figure 5.22 Representation of graph using pointers and list structures

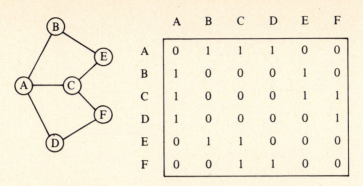

	A	B	C	D	E	F
A	0	1	1	1	0	0
B	1	0	0	0	1	0
C	1	0	0	0	1	1
D	1	0	0	0	0	1
E	0	1	1	0	0	0
F	0	0	1	1	0	0

Figure 5.23 A graph and its incidence matrix

Another method for storing a graph involves the construction of an *incidence matrix* (sometimes called a *connection matrix*). To form the matrix, list the nodes along the left side and across the top of the matrix. If there are n nodes in the graph, then there will be n rows and n columns in the incidence matrix or a total of n^2 elements. Place the number 1 in position i,j of the matrix (row i, column j) if there is an edge between node i and node j. If there is no edge between nodes i and j, place zero (0) in the matrix. Figure 5.23 illustrates a graph and its corresponding incidence matrix. For a non-directed graph the incidence matrix is *symmetric*, that is, entry i,j is the same as entry j,i for all values of i and j. For a directed graph the incidence matrix would not necessarily be symmetric. However, in defining the incidence matrix for a digraph we must be careful to denote direction. Thus the number 1 in position i,j means that there is an edge from i to j. If the edges are weighted, we could record the weights in the incidence matrix instead of using number 1. In the case of multiple edges between nodes, we could record the number of edges in the incidence matrix.

An interesting feature of the incidence matrix is that the number of paths of length two, three, four, and so on, can be calculated in a straightfoward manner. The original matrix M identifies those nodes that are joined by a path of length one (an edge). Nodes joined by a path of length two (two edges) are identified by the matrix M^2 (that is, $M \times M$). Paths of length three are in M^3 ($M \times M \times M$), and so forth. For example, consider the digraph in Figure 5.24. Listing all paths of length exactly two, we obtain $(v_1\ v_1\ v_2)$, $(v_1\ v_1\ v_1)$, $(v_1\ v_2\ v_3)$, $(v_3\ v_1\ v_2)$, $(v_3\ v_1\ v_1)$, and $(v_2\ v_3\ v_1)$. The square of the incidence matrix yields:

$$
M^2 = \begin{array}{c} \\ v_1 \\ v_2 \\ v_3 \end{array}
\begin{array}{ccc} v_1 & v_2 & v_3 \\ \hline 1 & 1 & 1 \\ 1 & 0 & 0 \\ 1 & 1 & 0 \end{array}
$$

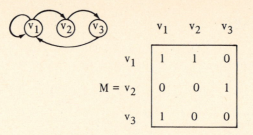

Figure 5.24 A digraph and its incidence matrix

Thus M^2 shows that it is possible to get from v_3 to v_2 by a path of length two, but not from v_2 to v_3 by such a path. Note, however, that M^2 does not say what the path is; it says only that the path exists.

Multiplying M^2 by M to produce M^3 we can determine those nodes connected by a path of length three. This yields:

$$M^3 = \begin{matrix} & v_1 & v_2 & v_3 \\ v_1 & 2 & 1 & 1 \\ v_2 & 1 & 1 & 0 \\ v_3 & 1 & 1 & 1 \end{matrix}$$

Even though our original matrix contained only 0s and 1s, we see that successive multiplications of the matrix by itself can introduce integers other than 0 or 1. This number indicates how many paths there are between two nodes. In this example we discover that there are two paths of length three between v_1 and itself. Saying this a different way, there are two cycles at v_1. Examining the graph we see that they are $(v_1\ v_1\ v_1\ v_1)$ and $(v_1\ v_2\ v_3\ v_1)$.

The preceding discussion suggests that there is no best way to store a graph. Often the application helps decide the way the graph should be represented. The most concise method is not necessarily the best. If one desires information about the various paths among nodes, the incidence matrix is very useful even though it takes more memory than other techniques.

Let us now consider a problem that frequently occurs in transportation or communication networks. Given a *network* (a weighted, connected graph), find a way to connect all the nodes so that the total weight is minimized. This means that the resulting structure has no cycles and is therefore a tree. Such a tree is called a *minimal spanning tree* because it connects (spans) all the nodes of the graph. The branches of the tree are edges in the original network. For example, Figure 5.25 shows a network and its minimal spanning tree. The weights are the lengths of the edges.

Algorithm MINSPAN is used to construct the minimal spanning tree of a network. It assumes that the weights of the edges are known.

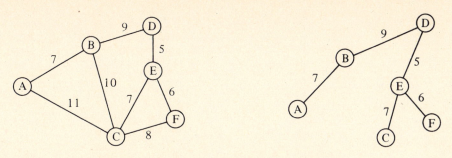

Figure 5.25 A network and its minimal spanning tree

Algorithm   MINSPAN

1. Comments: Find the minimal spanning tree of a network.
   Input: network
   Output: minimal spanning tree

2. Select any node and place it on the tree.

3. Repeat the following until the tree contains all the nodes:
   (a) From the nodes not in the tree select a node that has an edge of minimal weight and joins the node to one already in the tree.
   (b) Include this node and edge in the spanning tree.

4. End MINSPAN

Choosing the appropriate data structures for the network and its minimal spanning tree can simplify the implementation of the MINSPAN algorithm. For the network we choose an incidence matrix such as the one in Figure 5.23, and place the weights of the paths in the matrix instead of the 1's. For the minimal spanning tree, we also use an incidence matrix, since the minimal spanning tree is a subset of the network. Initially, finding the minimal weight in step 3(a) means that we look at only one row of the matrix—the row corresponding to the node selected in step 2. Later executions of step 3(a) require that we examine more rows of the matrix. For example, refer to Figure 5.26 and note that if we select node C first, the next node to add to the tree is E. Having selected C and E for the tree, we look at both their rows and find that the likely candidates for the next node are A, B, D, and F. The node with minimal weight (D) is added.

The following portion of a Pascal program gives an implementation of the MINSPAN algorithm. It assumes that the network is stored in an N × N incidence matrix named NETWORK. The network is an undirected graph; hence, the matrix is symmetric. The N × N matrix MSPTREE is initially zero, but it will become the minimal spanning tree. NODES_ON_TREE is a vector that tells whether or not a node is on the minimal spanning tree. If NODES_ON_TREE[I] is 0, then node I is not on the tree. If

	A	B	C	D	E	F
A	0	7	11	0	0	0
B	7	0	10	9	0	0
C	11	10	0	0	7	8
D	0	9	0	0	5	0
E	0	0	7	5	0	6
F	0	0	8	0	6	0

Figure 5.26 An incidence matrix for the graph in Figure 5.25

NODES_ON_TREE[I] is 1, node I is on the minimal spanning tree. To simplify the programming, we use numbers instead of letters for the nodes.

```
(* Explanation of variables:                                              *)
(*    NETWORK          An N × N array representing a network; values in    *)
(*                         the NETWORK are weights of paths connecting     *)
(*                         the nodes                                       *)
(*    N                Number of nodes in the NETWORK                      *)
(*    MSPTREE          Minimal spanning tree of the NETWORK                *)
(*    LENGTH           Number of nodes in MSPTREE                          *)
(*    MINVAL           Minimal weight of nodes not on tree                 *)
(*    MINNODE          Node number of node with minimal weight that is to  *)
(*                         be added to tree                                *)
(*    MINI             Node number to which node with minimal weight is    *)
(*                         connected                                       *)
(*    NODES_ON_TREE    Vector that tells whether node I is on tree or not; 0*)
(*                         = not on tree; 1 = on tree                      *)
(*    INFINTY          A large number; initial value of MINVAL             *)
(*                                                                         *)
    NODES_ON_TREE[1] := 1;          (* Put first node on tree;             *)
                                    (* Assume that we start with node 1    *)
    LENGTH := 1;                    (* One node on tree                    *)
    writeln (OUTPUT, 'ADDED NODE: ', 1);
repeat                              (* Loop until all nodes are on tree    *)
    for K := 1 to N do             (* Search for first node not on tree   *)
       if NODES_ON_TREE[K] = 0 then
         begin
            MINVAL   := INFINTY;    (* Set min = "a large number"          *)
            MINNODE := 0;           (* MINNODE will record the node        *)
                                    (* number of the node with minimum     *)
                                    (* weight                              *)
                for I := 1 to N do
                   begin
                      for J := I to N do
```

```
        if (NODES_ON_TREE[I] <> 0 ) &
                        (* Is node I on tree?                      *)
           (NODES_ON_TREE[J] = 0 ) &
                        (* Is node J not on tree?                  *)
           (NETWORK[I,J] <> 0) then
                        (* Are I and J connected?                  *)
        begin
                        (* Find minimum weight of nodes not on *)
                        (* tree, but connected to some node cur- *)
                        (* rently on tree                          *)
          if NETWORK[I,J] < MINVAL then
            begin
            MINVAL   := NETWORK[I,J];
            MINNODE := J;
                        (* Save node number                        *)
            MINI     := I;
                        (* Save connection node number             *)
            end;
          end;
        end;
                        (* add node to minimal spanning tree     *)
        MSPTREE[MINI,MINNODE]    := MINVAL;
        MSPTREE[MINNODE,MINI]    := MINVAL;
        NODES_ON_TREE[MINNODE] := 1;
        writeln (OUTPUT, 'ADDED NODE: ', MINNODE, ' CONNECTED
            TO ', MINI);
      LENGTH := LENGTH + 1;
    end;
until LENGTH >= N;                (* Stop when all nodes are on tree    *)
```

This program writes the nodes and their connections to the spanning tree as they are found. It also records the nodes in the minimal spanning tree. Since the original network is an undirected graph, the resulting minimal spanning tree is an undirected tree.

Finding the shortest path from one node to another is also a common problem in graph theory (see Chapter 10). It involves scanning a graph in such a way that all nodes of interest are considered and that no node is considered twice.

Like the algorithm for tree scan, a graph scan uses a stack to keep track of possible paths to try, but it also includes a flag with each node to tell whether or not the node has been previously scanned. Since graphs can have cycles, the VISIT flag will keep us from getting into an infinite loop. Algorithm GRAPH_SCAN assumes that it is given a digraph and a particular node. It will then scan the graph, visiting each node only once. (Algorithm GRAPH_SCAN can easily be modified to work for nondirected graphs as well as directed graphs.) The notation VISIT[N] refers to the value of VISIT for node N; each node has its own VISIT flag. If VISIT[N] = 1, the node has been visited. If VISIT[N] = 0, the node has not been visited.

Algorithm GRAPH_SCAN

1. Comments: Scan all the nodes of a digraph, visiting each node only once.
    Input: node   Where scan should begin
    Output: graph nodes In scanned order
    Variables: VISIT[I]   Array of flags denoting whether or not node I has been
                             visited; 0 = not visited; 1 = visited
                 N   Node being scanned

2. Mark all nodes as not visited:
       VISIT[I] ← 0 for all I in graph

3. Begin scan with input node:
       N ← input node

4. Repeat the following until stack underflows:
    (a) Place all nodes pointed to by N on stack.
    (b) Output node N.
    (c) Mark node N as visited:
            VISIT[N] ← 1
    (d) Remove nodes from stack until you find one that has not been visited. Set
         N ← next node to visit.

5. End GRAPH_SCAN

   Applying Algorithm GRAPH_SCAN to the digraph in Figure 5.27 yields one possible ordering: A, B, D, C, E. Will the GRAPH_SCAN algorithm give only one possible solution? Not necessarily. The order depends on the order in which the nodes are placed on the stack. For example, starting at node A, step 4(a) says to put B and C on the stack. If we put B on first and then C, we do not obtain the same scan order that we would if C went on the stack before B.

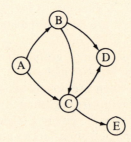

Figure 5.27 A directed graph to scan

EXERCISES

1. The figure below represents the graph introduced as a puzzle by Sir William
    Rowan Hamilton in 1859. Each node was named after a different city and the
    object was to find a simple path that visited each city only once. Begin at node
    1 and list the nodes and edges for such a path.

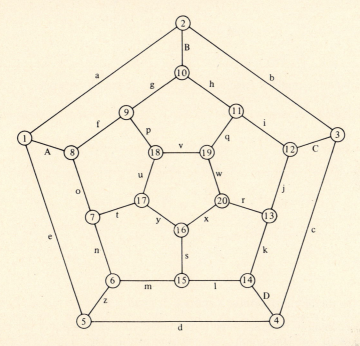

2. Compare the methods for storing graphs. Which one requires the most room in
    memory? Which one the least? Can you devise other methods for storing graphs?

3. Give both the tabular method (as illustrated in Figure 5.21) and the list-pointer
    method (Figure 5.22) for the digraph below.

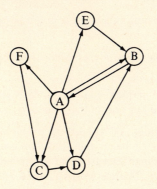

4. If a graph contains n nodes, what is its maximum number of edges? What is the maximum number of edges for a digraph with n nodes?

5. Given the following incidence matrix, draw its corresponding graph. Compute M^2 and M^3 for the graph.

$$M = \begin{bmatrix} 0 & 0 & 0 & 1 \\ 1 & 0 & 1 & 0 \\ 0 & 1 & 0 & 0 \\ 0 & 1 & 0 & 0 \end{bmatrix}$$

6. Determine the minimal spanning tree for the following network.

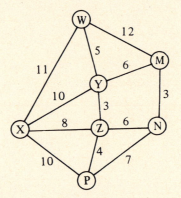

7. Scan the digraph in Exercise 3 starting at node F.

8. Write a computer program to perform the GRAPH_SCAN algorithm. What is the best way to store the graph? Can you improve the algorithm?

9. How do cycles show up in the incidence matrix for a graph? Be sure to consider the original matrix M and its powers M^2, M^3, and so on.

10. Is it possible to use threads to aid in the scanning of a graph? What are some of the problems involved?

11. Suggest alternate devices for scanning a graph so that all nodes will be scanned and infinite loops will be avoided.

12. Modify the Pascal implementation of Algorithm MINSPAN so that it works for directed graphs.

13. Incorporate the Pascal implementation of MINSPAN in a complete program supplying appropriate input and output procedures.

14. Draw a graph of the Bridges of Königsberg problem in Figure 5.28. This problem prompted Euler (1736) to begin the study of graph theory. The residents of Königsberg (now Kaliningrad) wished to promenade around town on Sundays, but they had a problem: The town was divided into four parts by a river. How could they cross each of the seven bridges one time only before returning to their

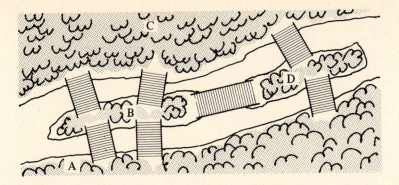

Figure 5.28 The Bridges of Königsberg connect land masses A, B, C, and D

starting point? (*Hint:* Let the land masses be nodes and the bridges be edges of the graph.)

15. Develop an algorithm that will find the minimal spanning tree of a graph stored as a linked list.

16. Design an algorithm to scan a graph stored as a linked list.

17. Write an algorithm to search a graph for a particular node.

6
FILE STRUCTURES

6.1
Some Hardware Considerations

In this section we will discuss basic hardware considerations that are pertinent to the applications in the remainder of the book. If you already know about peripheral devices and how they work, you may wish to skip this section.

Often the importance of hardware is underplayed in computer classes. Large computer centers have specially trained personnel who operate the equipment, so students may never have the chance to see computers in action. On the other hand, microcomputers are becoming prevalent in colleges, high schools, and homes, so more and more people have the chance to become familiar with hardware.

In many cases computer hardware imposes restrictions on the types of data structures that can be implemented efficiently. For example, complex structures, such as trees and graphs, are not easily manipulated if stored on magnetic tape. The volume of data, its frequency of access, the access keys, and amount of updates are also important in the choice of the appropriate storage medium.

Most computer systems contain two types of storage: main storage and auxiliary storage. The central processing unit can access main storage (main memory) directly, but must access auxiliary storage indirectly. Anything placed in auxiliary storage must first be transferred into main memory for processing. Typically, main memory is constructed of semiconductor circuits. Its size varies anywhere from less than a million to a few million characters.

Auxiliary (secondary) memory usually takes the form of rotating or moving magnetic material (disk, drum, or tape). It is slower and less expensive than main memory, but can store a much larger volume of data.

Locating data on auxiliary storage devices is facilitated by directories that specify the address of certain items and other pertinent information. These directories are used frequently by the operating system and consequently must be organized efficiently. Part of the structure of these directories depends on design features of the auxiliary memory. The operation of addressing data on different devices is particularly important. Before investigating the structure of the directories, let us discuss the two most common forms of auxiliary storage: magnetic tape and disk.

MAGNETIC TAPE

Magnetic tape is often used to handle large volumes of information. It is relatively inexpensive and keeps data close at hand so that it can be read into memory when required. It is also a good medium for transmitting large volumes of data from one physical location to another.

Standard magnetic tape is a strip of plastic $\frac{1}{2}$ inch wide, coated with a magnetic oxide material. A standard reel of tape 2400 feet long can hold up to 144 million characters.

The device that reads or writes information on the tape is called a *tape drive* (Figure 6.1, p. 124). The tape drive records data when a tape passes over an electromagnetic device called a *read/write head*. The head writes characters by magnetizing small areas on the oxide coating of the tape. Each character is a specific pattern of dots extending across the width of the tape. For example, in a seven-track tape, the pattern of dots requires seven positions. Each position is called a *track*. Figure 6.2 illustrates some of these patterns for seven- and nine-track tapes. A zero on a seven-track tape is represented by magnetized areas (dots in our figure) in tracks 4 and 6. A zero on a nine-track tape is represented by magnetized areas in tracks 3, 4, 5, 6, and 7.

The seven-track coding always represents a character as an even number of magnetized areas (bits). This is called *even parity*. Nine-track coding, on the other hand, uses *odd parity*. Each character is an odd number of bits. On a seven-track tape, six of the tracks are used for data. The seventh [labeled P in Figure 6.2(a)] is the parity track. The maximum number of characters that can be represented using six bits is thus 2^6, or 64. Nine-track tape uses eight tracks for character encoding and one track (actually track 4) for parity. The maximum number of characters that can be represented with eight tracks is 2^8, or 256.

In main storage, we typically work with a single number or character, because we can address it directly. On auxiliary storage devices, we work with records or atoms of data. Recall that a record (or atom) is a collection of related data items, and that the term *record* is typically used with auxiliary storage devices. We work with records for both logical and practical reasons. It takes time to read or write data from auxiliary storage. We can minimize that time by working with blocks of related items.

To write a record on a tape, the tape drive uses an induced magnetic field

Figure 6.1 Tape drive (Courtesy of IBM)

to magnetize the iron-oxide coating on the tape. Any previously recorded information is erased. Writing data takes place as the tape moves past the read/write head on the tape drive. First the tape accelerates from a stopped position to its operating speed and then slows down and stops. As the acceleration and deceleration take place, some tape passes by the read/write head and remains blank. Thus between every record of data on the tape there is a blank area. This area is called an *inter-record gap* (IRG) or *inter-block gap* (IBG), since records are frequently grouped together into a block. The size of this gap varies from $\frac{3}{10}$ to $\frac{3}{4}$ inch, depending on the tape drive.

A *tape mark* is a special character that can be written on the tape to denote the end of data. The tape mark is also known as an *end-of-file mark*. Since the end-of-file mark may not appear at the physical end of the tape, we may be able to write several files on the same reel of tape. For example, we may want to save on tape copies of our debugged programs, so that we can retrieve

(a) Seven-track tape

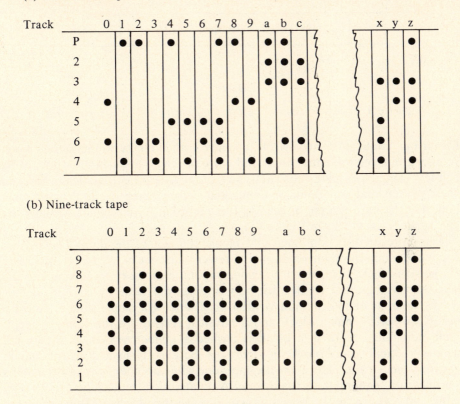

(b) Nine-track tape

Figure 6.2 Representation of characters on magnetic tape

them as necessary at a later time. Each program would be a separate file on the tape.

The physical beginning and end of the tape is marked by a small metallic strip. At the beginning of the tape, it is a *load-point* marker. At the end, it is an *end-of-reel* marker. All data are written between the load point and the end of reel. When a magnetic tape is loaded on a tape drive, it should be positioned at its load point. We can then issue instructions in our programs to read or write the tape, to scan forward to the end-of-file mark, to rewind the tape back to its beginning, or some other such operations.

The number of characters that can be stored per inch of tape defines the tape's density. Common tape densities are 800, 1600, and 6250 bpi (bytes per inch). At 6250 bpi, 80 characters occupy less than .013 inch. At a density of 6250 bpi, the IBG is $\frac{3}{10}$ inch. So if information is written in 80 character records, most of the tape will be blank (see Figure 6.3).

This may be an extreme example, but it illustrates how careful we should be about choosing the size of our records and blocks written to tape. Even

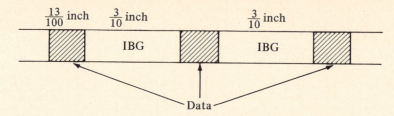

Figure 6.3 *Representation of 6250-bpi tape with 80-character records*

though our application may work logically with records 80 characters long, we may group records into blocks and write or read one block at a time. Most high-level languages have constructs, so that records are automatically blocked when written and separated or deblocked before being used. Programmers think that they are writing or reading one record at a time, when actually the tape appears as shown in Figure 6.4.

The more records contained in a block, the denser the data. That is, the tape will contain more information and fewer blank areas. However, to create a large-size block, there must be enough room in memory to accumulate the data. All the data in a block must be written at the same time.

The amount of information written on a tape at a single time is called a *physical record*. Thus a block and a physical record are the same concept. Each individual record in a block is termed a *logical record*. In Figure 6.3 there is 1 logical record/block. In Figure 6.4 there are 50 logical records/block.

Since blocks are placed on the tape one after another, we say that tape is a linear or sequential access device. Any additions to the tape can go only at the end. Otherwise the new data would erase the old data. This sequential nature also makes it difficult to delete a record or to search for a particular record. In fact, working with a tape is identical to working with a dense-list structure. Because of these problems, other types of secondary storage have been developed.

Rotating memory devices, such as magnetic drums and disks, are similar to tape in some ways, but allow direct access to any physical record without processing all the preceding records. A disk is similar to a thin, metal phonograph record, but it is magnetically coated. Several disks are mounted together on a vertical shaft to constitute a disk pack. The number of disks in a pack varies from model to model and manufacturer to manufacturer. Each

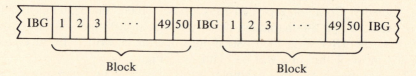

Figure 6.4 *A tape with fifty 80-character records per block*

Tracks 000–200

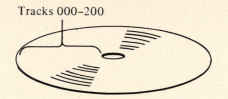

Figure 6.5 Diagram of tracks on a disk

surface of a disk contains concentric tracks (not tracks in a spiral, as on a phonograph record) wherein the data are magnetically encoded. The number of tracks per surface varies with the disk model and manufacturer. Figure 6.5 illustrates a single disk with 200 tracks/surface.

In a movable-head disk device the read/write heads are located on arms that move in and out over the disk to reach a specified track, while the disks rotate beneath the heads (see Figure 6.6). The heads fly on a cushion of air and do not actually touch the surface of the disk. All the disk's arms move simultaneously. All the heads are positioned over the same track, but on different surfaces. These vertical tracks constitute a cylinder. Even though only one head can read or write at a time, the other heads are ready to access other data on the cylinder.

Data are placed on each track in records. Records on a track are numbered sequentially from a location called the *home address* (track address). Records

Read/Write Heads

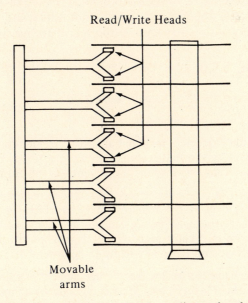

Movable
arms

Figure 6.6 Diagram of disk read/write heads

are separated by gaps, much like the gaps on tape, and overflow is handled by placing records on the next lower track of a cylinder.

To access a record in either a read or write operation, three steps are involved:

1. Seek: Move the read/write heads to a position over the proper track.
2. Search: Activate the appropriate head and locate the desired record on the track by counting from the home address.
3. I/O: Read or write the record.

Often the operating system or disk hardware mechanisms take care of these steps, so the programmer is not aware of them. However, two different time delays are involved: rotational delay (looking for the home address) and seek delay (movement of the read/write heads). These delays are on the order of 10^{-2} second, while main memory access time is on the order of 10^{-7} second. Hence, disk systems access data at a much slower rate than main memory does. These access times may sound minute and the differences may seem insignificant. However, if these operations are repeated thousands of times, the differences become important.

A *drum* is sometimes called a fixed-head disk, since it has a read/write head for each track. It offers faster access to the data because it requires no seek operation, as a disk does. However, the cost of a drum typically is more than that of a disk, and the storage capacity of a drum is less than that of a disk. For these reasons, a drum is frequently reserved for portions of the operating system that require fast access, but which are either too large or are used too infrequently to remain in the main storage area.

EXERCISES

1. Why are main and auxiliary memories needed?

2. What causes gaps on magnetic tape?

3. If a record is 100 characters long and an IRG is $\frac{3}{10}$ inch long, how much of the tape will be blank, given that records are blocked 10 per block? 20 per block? Perform these calculations for tapes with density 1600 bpi and 6250 bpi.

4. A random search of a tape for a particular record requires that nearly half the tape be read. A disk search requires on the average that half a track rotate beneath the read/write head before locating the home address. Why do these operations require one-half a scan? (*Hint:* This is an average computed by assuming that each record is equally likely to be accessed.)

5. For a disk containing 200 tracks, the average number of cylinders crossed during a seek operation is 67 ($\frac{1}{3}$ times 200). How is this average computed? [*Hint:* Assume

that each cylinder is equally likely to be accessed. Also, the average of the distribution of the absolute difference of two random numbers ($k = |i - j|$) is $\frac{1}{3}$.]

6. Write a program that will read data from tape and will print every one-hundredth record.

7. Take a program that normally prints its output and modify it so that it writes the data to a disk file. How can we be certain that the program is working properly?

8. Investigate the hardware at your computer center. How much storage does the computer have? What is the capacity and speed of the disk drives? Are the tape drives seven- or nine-track? What other facts can you learn about the hardware?

9. Give examples of applications that would use tape for input and/or output.

10. Give examples of applications that would use disk for input and/or output.

11. Write a program that will take an input tape or input from disk and reblock the data onto another tape or disk file. Choose the block size that is appropriate for the tape or disk you have available.

6.2
Files

From the point of view of graph theory or data structure, files are nothing more than data structures that reside physically on secondary storage devices, such as disk or tape. File structures may be thought of as special data structures with an emphasis on retrieval efficiency. Since secondary storage devices are slower than main memory, a high penalty is paid for every access to information stored in a file. The objective, then, is to design a data structure for secondary storage devices that minimizes operating overhead. Such a design is based on the following considerations:

1. Volume of information

2. Frequency of retrieval

3. Frequency of update

4. Number of access key fields per record

The file designer must evaluate these considerations before deciding which structure to employ. After selecting a structure, algorithms for carrying out the needed operations must be programmed. These algorithms allow the user to store, maintain, and retrieve a set of data.

Historically, files have been studied separately from data structures. As a result, there is a distinct vocabulary for files, even though concepts in the two areas may be the same. Table 6.1 lists corresponding terms for data structures and files.

Table 6.1   Corresponding Vocabulary for Files and Data
Structures

Files	Data structures
Field	Field
Record	Atom or node
Key	Identifier field
Address	Pointer or address
File	Data structure (list, string, tree, graph)
Library	Collection of data structures

A *field* is a unit of information. A *record* is a collection of related fields treated as a unit. A *file* (or *data set*) is a collection of related records. For example, the entire collection of an insurance company's data would be a file. It might also be classified as a *data base,* since it is a collection of data that is a fundamental part of an enterprise. The notion of a data base frequently includes the concept of shared data—that which can be accessed by many users. Not everyone may have access to all the data, since it may not be relevant or pertinent to his or her job. The topic of data base encompasses too large a subject area to be included in this text. Therefore, we omit discussion of the topic, except where its concepts are also related to data structure.

To illustrate the definition of data structures for files, let us consider an insurance company's policy file. The policy number, premium rate, and expiration date are examples of fields. Information pertaining to a single policy is a record. To facilitate handling of data, records are formatted so that each field occupies a particular location within the record. In addition, field lengths are kept constant. For example, Figure 6.7 illustrates a record format for an insurance company. The record designer must decide not only what to include in a record but also the length of each field. If the name field is limited to 30 characters, there must be a rule for handling a name with more than 30 characters. For ease in reference, the name and address fields might be divided into subfields of last name, first name, middle initial, street, city, state, and zip code. If we divide into subfields, we must decide the exact location of each subfield. Thus, if we are searching for policyholders in Texas, we do not have to scan the entire 50-character address field looking for TX.

Policy number	Name	Address	Coverage codes	Premium	Expiration date	Last use

1   10 11   40 41   90 91   100 101   105 106   111 112 117

Figure 6.7 Format of record for insurance file

A *key* is one or more characters of a record used to identify the record or to control its use. Typically, a key is one of the record's fields. When the file is sorted, the key determines the order in which the records are arranged. Files may have single keys or multiple keys. A *single key file* has one key per record. A *multiple key file* has more than one key per record, but all records have the same number of keys. In a single key file, retrieval and update are solved relatively easily by careful selection of file structure. For multiple key files, file organization becomes a challenging problem.

Example   A registrar's file would normally contain a single key—the student's Social Security number—since there would be no chance of two students having the same number. However, an insurance company file might need two keys: the policyholder's name and account number. The account number is of primary importance to the company—but the name is naturally more significant to the policyholder.

EXERCISES

1. List the four most important parameters to consider in file structure design.

2. Give an example of the following structures: (a) Large file, (b) frequently updated file, (c) multiple-key file, (d) frequently accessed file.

3. Define the following terms and give examples: (a) Variable-length record, (b) key, (c) record format.

6.3
File Storage
and Retrieval

The storage of a file and the subsequent retrieval of its records go hand in hand. The storage method determines the manner and speed of accessing the data. Basically, there are two types of access modes: sequential and random. In a *sequential access file,* records must be processed in the order in which they were written on the file. For example, a tape file is a sequential file because of the nature of magnetic tape. Disk files can be either sequential or random access. In a *random access file,* records can be retrieved in any order either because the position of a record is known or it can be determined from a key. Many file organizations are of the random access type. We will discuss the basic ones in this section. First, however, let us consider the simplest organization, that is, the sequential.

A sequential file is a good system for handling a large volume of data that rarely changes and is processed as an entire unit. A company's payroll for salaried employees is a good example. Every payday the entire file is read and checks are printed. If raises are granted, new employees hired, or old employees retired, a file update is necessary. If the file's record order is not important, new records can be added to the end. But how can changes or deletions be handled? If the changed record is the same size as the previous record, it may be possible to write the new record over the old. On a tape file, writing new records over old records may not work unless the tape drive is adjusted properly. (Remember that tape drives start and stop. A slight differential in the operating speed could place the inter-record gap in the wrong place, making it impossible to access other records on the tape. See Figure 6.8.) In Figure 6.8 the dashed lines show misalignment of data, so part of data$_2$ is taken up by IRG. If the file's records are blocked, a change to a single record means that the entire block must be rewritten.

A sequential file is sometimes updated by making a new copy. With tapes, this means writing a second tape, copying the unchanged records from the original file, deleting (not writing) specified records, and inserting new records where required (see Exercise 6). The old tape can be kept as a backup or reused for another file. With disk files, the operations are similar. After the updated copy is created, the previous disk space is available to other users. Thus requirements of disk management and memory management are similar.

If the sequential file is ordered according to a key field, updating necessitates creating a new file. Typically, all records to be updated are placed on a separate file and are ordered in the same way as the original master file. Then the master file and update file are merged to form a new master file. Keeping both the master and update files in order facilitates the merger.

Example Merger of master file and update file to create a new master sequential file. The file contains a warehouse inventory, including part number, number on hand, and reorder point. The file is arranged according to part number. Each record of the update file contains a code that tells whether the record is an addition to the inventory (A), a change to the inventory (C), or a deletion (D).

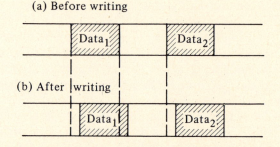

(a) Before writing

(b) After writing

Figure 6.8 Misalignment of data during a rewrite

OLD MASTER FILE			UPDATE FILE			
Part number	*On hand*	*Reorder point*	*Part number*	*On hand*	*Reorder point*	*Code*
1076	460	200	1091	120	85	A
1088	240	100	1098	45	20	A
1093	25	50	1104			D
1104	362	300	1152	950	500	C
1152	891	500				
1683	208	120				

To create the new master file, we must read both the old master and update files, reading one record at a time from each file and comparing the records. Initially, the master file part numbers are less than the first update record (1091). This means that all records from the master file can be copied directly to the new file until a part number equal to or greater than 1091 appears. When 1093 is read from the old master, we know that 1091 must be an addition to the new master file. After adding 1091, we go on to the next update record (1098) and continue copying old master records until we can correctly insert 1098. The next record is 1104, which we delete. We continue to process in this manner to the ends of both files. For more information, consult B. Dwyer and J. Inglis in the Reference section.

Example   The regional office of a medical insurance company wants to design a file containing information on its current policies. Each record will consist of policy number, policyholder's name, type of coverage, premium, expiration date, type of last claim, and settlement. The purpose of the file organization is to provide easy access for hospitals. When a person enters a hospital, a telephone call will verify the patient's insurance coverage and obviate the need for the usual hospital written forms.

All these considerations mean that, once it is created, the file will be large, retrieval must be rapid (while the hospital waits on the phone), and updates will occur less often than retrievals. The search of a sequential file for a particular record usually requires scanning half the file. (This fact is demonstrated in Chapter 8.) Therefore, the best way to organize the file is to use one of the random access methods. As you read about the various access methods, keep this problem in mind and determine the best method for the insurance file.

DIRECT ACCESS

One random access method is *direct* access. With this type of organization there is a special relationship between a record's key and the key's actual address. Each key has a corresponding unique address, so that only one seek and one read are necessary to retrieve the record. Of course the key-address relationship depends on the data in the file and the storage device chosen for the file. As a very simple example, suppose that each record in a file occupies one track and that the records are numbered 0001, 0002, and so on. Assume further that the disk has 10 tracks/cylinder. To make efficient use of the disk's head movement, records from 1 to 10 are on the same cylinder, 11 to 20 on the next cylinder, and so on. Given record number n, we thus divide by 10 to compute the cylinder number (relative to the beginning cylinder allocated to the file) and use the remainder of the division to specify which disk surface or head the record is on. For example, if the file begins on cylinder 25 and ends on cylinder 40, record 0043 is on cylinder 29, head 3.

$$\text{Cylinder} = \text{beginning cylinder} + \frac{\text{record} - 1}{10}$$

$$= 25 + \frac{42}{10} = 29$$

$$\text{Head} = \text{remainder of } \frac{\text{record} - 1}{10} + 1$$

$$= \text{remainder of } \frac{42}{10} + 1 = 3$$

INDEX (Dictionary)

To avoid computing a record's location, we can keep an index (dictionary) of all key values and corresponding addresses. Retrieving a record involves scanning the index for the record's key, getting the key's address, and reading the record. Retrieval time is shortened because only one access is made to the secondary storage device. The index requires space and search time, but searching the index may be cheaper than searching records stored in secondary memory.

Updates to the file involve updating the index. If a record is deleted from an auxiliary storage device, it need not be altered as long as its entry is deleted from the index. If there are many deletions, we would want to recover the space belonging to the deleted records and make it available for reuse. If a record is added to the file, its physical location in relation to other file records is not important. The important aspect is the inclusion of the record address and its key in the index.

If the file is large, the index can become a problem. The size of the index depends on the number of records in the file. As index length increases, search time also increases. We can solve this problem by using a sequential organization for the file with an index to certain key points in the file.

Chapter 10 discusses two methods of indexing a file so that a single record can be retrieved without much searching of the file. One method is a B-tree index organization (Section 10.10); the other is extendible hashing (Section 10.11). You may wish to study these methods at this time, since they are pertinent to the study of files. Each section in Chapter 10 is self-contained, so it can be read whenever appropriate.

INDEXED SEQUENTIAL

The indexed sequential file requires the file records to be ordered according to a key. Rather than scan the entire file for a particular record, we consult a partial index that indicates approximately where to start and how far to continue scanning before deciding that the record is not in the file. The partial index is set up when the file is created, according to the way records are grouped into blocks. There is one entry in the index for each block. It contains the location of the block and either the lowest or highest key (depending on the ordering) in the block. If the file is large, even the partial index may be unmanageable. In this case we can define a higher-level index that does the same thing for the partial index that the partial index does for the file.

For a simple example, consider the parts inventory file. The records are referenced by part number and the numbers are in increasing order. The file is stored on cylinders 21 through 25 and occupies 10 surfaces on each cylinder. The high-level index gives the location (cylinder number) of the lower-level indexes. The part number represents the largest part number stored on that cylinder.

High-level Cylinder Index

Cylinder	Part number
21	0829
22	1450
23	2079
24	3635
25	4721

Each cylinder has its own low-level track index, which gives the surface number and highest part number stored on that surface. For example, the track index of cylinder 23 might look like the following.

Low-level Track Index for Cylinder 23

Surface	Part number
1	1512
2	1582
3	1640
4	1690
5	1725
6	1776
7	1834
8	1899
9	1963
10	2079

The cylinder index helps direct us to the correct track. For example, any part with a number between 1451 and 2079 will be on cylinder 23, if it is in the file. The cylinder index tells us that no part numbers above 4721 exist in the file. To retrieve information on part number 1703, we must first search the cylinder index to determine the correct cylinder (23), and then search the track index of cylinder 23 to locate surface 5. Then cylinder 23, surface 5 must be read and each record's key compared with 1703. One of two things will happen: Record 1703 will be found and its information will be copied into main memory for processing, or a key larger than 1703 will be read, implying that there is no part number 1703.

Remember that an indexed sequential file is a sequential file and its records can be processed sequentially. On the other hand, partial indexes permit relatively fast random access. Thus it is a flexible file organization. However, insertions and deletions may necessitate updating indexes and may create overflow areas, which slow down access to individual records.

TREE

The tree organization is used frequently to define file directories and to determine access rights and privileges. Most large computer systems allow users to define their own disk files. Each user can declare the file sharable or nonsharable, read-only, write-only, read-and-write, and so on. Some systems allow users to specify those who have access to a file and their corresponding access rights. For example, some users may have read-only privileges, others read-and-write. Protecting the files from incorrect access is an important problem in large file or data-base systems that do permit sharing of data.

A tree is a good way to represent the file directory. The nodes of the tree can contain the access information and location of the file. Thus the operating system can check the directory to see if a user has access to a particular file, and can then locate the file. To illustrate the tree organization of the directory, we will assume that there are two basic types of files: system files and user-created files. The *system files* are compilers, assemblers, and various other utility programs that anyone can use (on a read-only basis). The *user files* are those created by the users. System files are normally referenced by name, user files by name and/or user number. Figure 6.9 gives an example of the directory tree.

In Figure 6.9 user files are classified under the number of the user who created the file. If user M4613 wanted to access the CEN file, the name CEN would have to be used. Most systems look first in a user's set of files before seeing if the request is for a system file. Thus the system would proceed down the user file branch to user number M4613 and then would retrieve the CEN file. If the same user requested PROC, the system would not find it under M4613 and would go back to the directory root and search the system file branch. If M4613 requested *B4810.PROC, the system would look under user

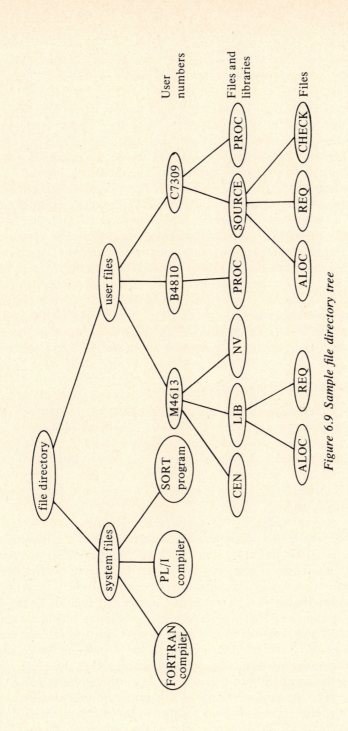

Figure 6.9 Sample file directory tree

B4810 for PROC. The B4810. preceding the name PROC qualifies PROC and says that it is in a special part of the file tree. The asterisk (*) is used in our example to denote that B4810 is a user number. Users can also establish libraries (collections of files) and reference a particular file in the form: library-name.file-name.

Thus user C7309 would access REQ with the phrase SOURCE.REQ. The name REQ would not be good enough because the system would look for a name on the same level as SOURCE. The methods of qualifying and identifying user number make it possible for two users to use the same file names. Yet the system keeps them separate because of the directory.

An advantage of the tree organization is that deletions and insertions can be made without affecting the rest of the structure. For example, if new users create files, we add their numbers and file names to the user file subtree. If users erase all their files, we simply remove their branches of the tree. As with all tree structures, insertions and deletions require extra memory space for needed pointers, but the technique permits fast access.

Figure 6.9 is only a diagram of the tree structure. Figure 6.10 gives an implementation of the tree directory. Each node of the tree contains two pointers: One references the next node at the same level in the tree and the other locates the node's successors. Although not included in the diagram, terminal nodes would contain information about the file and its protection.

RANDOMIZING (Hashing)

Hashing is a searching technique that attempts to locate the desired item after one trial. It applies equally well to searching for items in main or auxiliary storage. In this section we introduce the topic of hashing, but discuss it in more detail in Section 8.3. A form of hashing that applies directly to file access is extendible hashing (Section 10.11). You have the choice of postponing the study of hashing until Chapter 8 or of previewing it here and perhaps going on to extendible hashing.

The idea behind hashing is to perform a calculation on a key field and obtain a value which becomes the address (or index) of the record to which the key belongs. It is desirable that every key field compute (or hash) to a different address. Then each record can be retrieved directly after performing the calculation. Hashing or randomizing is generally used if the entire range of keys is large compared with the actual number of records in the file. For example, in a company that uses a six-digit employee number, there are a million possible numbers, 000000 through 999999. Instead of reserving space for a million records, we perform a computation on the employee number to convert it to a range of 0 through 9999. Assuming that there are only 8000 employees in the company, we can say that our scheme looks good. To store a record, we take the key, perform the computation, and obtain an address. We then store the record at that location. This scheme is also called a

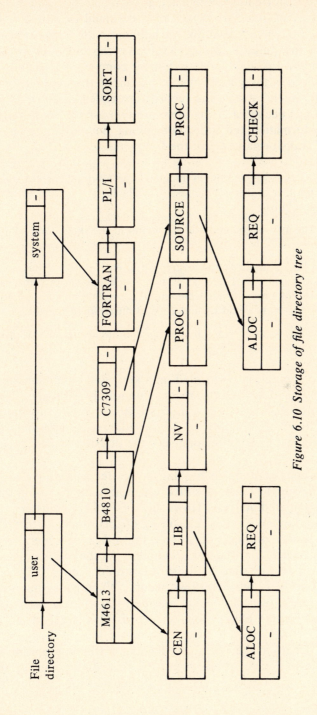

Figure 6.10 Storage of file directory tree

key-to-address transformation. When accessing a record, we perform the same computation and determine where the record should be stored.

For example, to convert an employee number in the range 0 through 999999 to a number in the range 0 through 9999, we can either drop the last two digits, drop the first two digits, or use a more complicated scheme. However, no matter what we do, there is always the likelihood that two keys will produce the same result. In this event, the records are called *synonyms,* and we say that a *collision* has occurred. Thus there are two aspects to hashing: the key-to-address transformation and collision handling. Several of the more common key-to-address transformations are the following:

1. Division: The key is divided by an integer (often a prime) slightly smaller than or equal to the number of distinct addresses available. The remainder of the division is used as the address.

2. Midsquare: The key is squared and the digits in the middle are retained for the address.

3. Folding: The key is divided into several parts, each of which has no more digits than the desired address. Then the parts are operated on in some way. Folding offers a variety of randomizing techniques, depending on the operation chosen. For example, the parts could be added or exclusive-ored together.

4. Random: The key is used as input to a random number generator. The output of the generator will be the address.

Ideally the computation performed on a key would yield unique addresses for different keys and scatter the records uniformly throughout the address space. Because of this, the term *scatter storage* is applied to the technique, as well as randomizing, hashing, or key-to-address transformation. Surprisingly, one of the best hashing techniques—the division technique—is also the simplest.

The second consideration in hashing or scatter storage is handling collisions. One of the simplest schemes involves storing the record in the first free area after the computed address. However, this tends to cluster records with the same hashed key. With files, clustering of records leads to the definition of buckets. A *bucket* is a unit of storage that can hold several records. On a disk a bucket might be a track or a cylinder. As the file is created, records whose keys hash to the same address are stored sequentially in the same bucket. If the bucket is full, overflow occurs. When accessing a record, we compute the address of a file bucket and perform a linear search (discussed in Chapter 8) within the bucket for the record. This is called the *open-hash* technique.

If overflow occurs in the open-hash technique (the appropriate file bucket is full), we have to decide where to put the record. We can, of course, use a nearby bucket, or a bucket specifically designated for overflow records.

The *loading factor* is a measure of the density of the file storage area. It is the ratio of the number of records to the total number of record slots in the primary file area, assuming that collision records are placed in the overflow file area. If each bucket has room for k records and there are b buckets in the primary storage area, the loading factor l is:

$$l = \frac{n}{bk}$$

where n is the number of records in the file. In general, the percentage of overflow records increases with a higher loading factor and decreases with larger bucket size. A low loading factor means a large number of vacant spaces in the primary storage area, but good retrieval time. Larger loading factors require a larger number of accesses to retrieve a record, but save total memory. For a particular file we must consider storage costs versus frequency of retrievals in deciding what the loading factor should be. For further applications of hashing, see Sections 8.3 and 10.11.

EXERCISES

1. Define the following terms:
    (a) Sequential file, (b) direct access, (c) hashing, (d) synonym, (e) bucket, (f) loading factor, (g) index.

2. Give one advantage and one disadvantage of each of the following file structures: sequential, indexed, indexed sequential, and tree.

3. Why is hashing difficult to apply to a multiple-key file?

4. A hashing function is to be applied to a file with a capacity of $k = 10$ records/ bucket. How many buckets are needed to store $n = 1000$ records and at the same time maintain a loading factor of $l = .85$?

5. Devise an update algorithm for inserting a new record in the indexed sequential file shown in the text. What are the problems encountered?

6. Devise an algorithm to update a sequential file stored on tape. Assume that tape 1 contains records to be inserted and/or deleted. Tape 2 contains the master file to be updated. The output of the update should go on tape 3.

7. How are bucket overflows handled in a file system employing buckets?

8. Describe how changes to a text file could be kept separate from the original source, but yet merged in before printing the text. What information needs to be kept in the change file? If the updated file were acceptable, then the changes could be made to the original.

7
SORTING

7.1
Simple
Sorts

Scatter storage and the techniques of multilinked lists might give the impression that the arrangement of data is not important. Indeed, the physical order of data within memory is not important, but for effective use of the data, it must be placed in logical order before it is of real value. Moreover, searching techniques (Chapter 8) are facilitated if data records are kept in some kind of order. As with most problems involving large data bases, sorting is best done by computer, using the more efficient sorting algorithms.

Sorting techniques fall into certain classes. Algorithms may be classified as *internal* or *external,* depending on whether the data being examined will fit in main memory or will require secondary storage. External sorting usually depends on the electromechanical capabilities of the secondary storage device, and incorporates an internal sorting algorithm as part of a user's program.

Algorithms may also be classified as comparative sorts or distributive sorts. *Comparative sorts* operate by comparing and exchanging two keys at a time. *Distributive sorts* separate keys into subsets so that all items in one subset are greater than all those in the other subset. Radix sort is an example of a distributive sort. Bubble sort is an example of a comparative sort.

Perhaps the simplest algorithm is the *bubble sort.* It compares successive pairs of elements in the list and interchanges two elements if necessary. The list is scanned repeatedly until no further exchanges are required: then the list is in order. Figure 7.1 shows how an item is "bubbled" to the top of the list. Comparisons begin at the bottom and work up. If there are n items in the list, the first pass requires $n - 1$ comparisons. Note that after the first pass, the smallest number is at the top of the list. After the second pass (which requires only $n - 2$ comparisons), the first two numbers are in their correct places. Sorting the entire list may require as many as $n(n - 1)/2$ comparisons.

Initial list	After first compare	After second compare	After third compare	After first pass	After second pass
19	19	19	19	01	01
01	01	01	01	19	11
26	26	26	26	11	19
43	43	43	43	26	21
92	92	92	92 · · ·	43 · · ·	26
87	87	87	87	92	43
21	21	21 ⟶ 11	11	87	92
38	38 ⟶ 11	11 ⟶ 21	21	21	87
55 ⟶ 11	11 ⟶ 38	38	38	38	38
11 ⟶ 55	55	55	55	55	55

Figure 7.1 Partial result of bubble sort

Another sorting technique that is easy to understand is the *selection sort*. Given a list of n elements, compare the first entry to the second, third, fourth, and so on. Any time that a value larger than the first value is found, swap it with the first entry. After going through the list once, we find that the largest value will be at the head of the list. Then compare the second element with the third, fourth, fifth, and so on. Again, swap elements so that the largest of the remaining $n - 1$ values winds up second in the list. Continue in this manner until the $(n - 1)$st element is finally compared with the nth, and the larger of the two is put in the $(n - 1)$st position. At this point the list is in order from high to low. A minor change in the algorithm would produce an ascending order.

These two algorithms are straightforward approaches to sorting and are easy to program. They are fine for short lists or for lists that do not need reordering often. However, these algorithms are too slow when frequent use is desired or when large amounts of data are involved.

To determine the "best" algorithm, we must establish standards. Two common measures are storage requirements and the average number of comparisons. We are posing two questions: What algorithms require the least amount of comparisons (on the average)? What algorithms require the least amount of storage?

Concerning the question of comparisons, what is the minimal number of comparisons possible? If a list is in order, it still requires a minimal number of comparisons to discover that fact. But if the list is in reverse order, we might expect that more comparisons will be required. The number of comparisons needed to order an unordered list is called the *sort effort*. The minimum sort effort for comparative sorts is on the order of

$$n \log_2 n$$

where n is the length of the list. For example, if the list contains 128 entries, then the minimum sort effort is approximately

$$128 \log_2 128 = 128 \cdot 7 = 896$$

This is a theoretical limit. Most algorithms require more comparisons. In the final analysis, the best algorithm is the one that runs fastest on a particular machine.

Other important things to consider for sort effort are the implementation of the algorithm and the machine on which it runs. It is possible that a good algorithm may have an inefficient implementation. The program may be easy to understand, may model the algorithm completely and correctly, and yet may not be as fast as other versions. If our primary consideration is speed, then we may have to program using lots of tricks and shortcuts that are difficult to decipher and comprehend when we return to study the algorithm later. In a learning situation we prefer readability and clarity. In a practical situation we must decide whether to sacrifice readability and clarity for speed.

The problem of analyzing algorithms is a very difficult one. Unfortunately there is no one sort algorithm that is good for every situation. Many of the algorithms have subtle variations designed to decrease the amount of storage required or to decrease the number of times we must examine the data. In this chapter we present several well-known algorithms and introduce a methodology for comparing algorithms.

EXERCISES

1. Perform a selection sort on the list in Figure 7.1.

2. What is the minimal sort effect for a list of 1000 items?

3. Give examples of distributive and comparative sorts.

4. Define the following terms: (a) Internal sort, (b) external sort.

5. Write an algorithm for a bubble sort. What data structure is used to store the values being sorted?

6. Write an algorithm for the selection sort.

7.2
More Sorting
Algorithms

The following algorithms are presented by example. The sorting effort is given without derivation and should be compared with the theoretical limit of $n \log_2 n$.

RADIX

The radix sort is the principle employed in mechanical card-sorting machines, which were developed years before computers. To sort numbers the algorithm requires the establishment of 10 queues, one for each of the digits 0 through 9. The algorithm examines the numbers one digit at a time, starting with the least significant (rightmost) digit. The radix algorithm follows.

Algorithm RADIX_SORT

1. Comments: Perform a radix sort for a list of numbers. Each number has K digits.
    Input: NUM   A list of numbers to be ordered
    Output: NUM   A list of ordered numbers
    Variables: K   Number of digits in each number
                 QUEUE0,   A queue for each of the digits 0 through 9
                 . . . ,
                 QUEUE9

2. Repeat the following for digit I, where I begins with the least significant digit and moves left one digit at a time toward the most significant digit.
    (a) For each NUM in the list, place NUM in QUEUEJ, where J is the Ith digit in NUM.
    (b) Reconstruct the list of NUM by concatenating QUEUE0, QUEUE1, . . . , QUEUE9.

3. End RADIX_SORT

The number of passes through the list equals the number of digits in the largest key. In Figure 7.2 the sort requires two passes because the largest number contains two digits.

It may seem that the radix algorithm examines the digits in the wrong order. It starts with the rightmost digit and works to the left. Use the numbers in the example and work from the leftmost digit to the right and see where the algorithm fails.

In general, for n keys, q queues, and k digits in each number, the storage S and the number of compares C are computed by averaging over uniformly occurring values:

$$S = (n + 1)q \qquad C = n \log_q k$$

Thus, in the example of Figure 7.2,

$$S = (16 + 1)10 = 170 \qquad C = 16 \log_{10} 2 \approx 5.0$$

A *binary radix* sort is sometimes preferred for binary machines, in which case the numbers are encoded into binary values and there are only two

Initial list	Pass 1 queues	First reconstruction	Pass 2 queues	Final reconstruction
19	0 70	70	0 01	01
01	1 21,11,21,01	01	1 19,11	11
26	2 92	21	2 26,21,21	19
43	3 43	11	3 38,36	21
92	4 54,64	21	4 43	21
87	5 55	92	5 55,54	26
21	6 36,26	43	6 64	36
38	7 77,87	64	7 77,70	38
11	8 38	54	8 87	43
55	9 19	55	9 92	54
21		26		55
64		36		64
54		87		70
70		77		77
36		38		87
77		19		92

Figure 7.2 Steps in radix sort for decimal keys

queues involved. For a binary radix sort of the numbers in Figure 7.2, we compute

$$S = (16 + 1)2 = 34 \qquad C = 16 \log_2 7 \approx 47$$

Thus an obvious tradeoff between storage and time is possible by changing the base of the numbers.

Note that the number of compares (C) for a radix sort may be less than the theoretical minimum of $n \log_2 n$. This is possible because the theoretical minimum applies only to comparative sorts. The radix sort is a distributive sort. It trades storage for speed.

BUBBLE

Bubble sort was discussed in Section 7.1. The maximum number of comparisons occurs when the items are in reverse order. Thus for a list of length n, the maximum is

$$C_{max} = (n - 1) + (n - 2) + (n - 3) + \cdots + 2 + 1 = \frac{n(n - 1)}{2}$$

The minimum number of comparisons occurs when the list is already ordered and

$$C_{min} = n - 1$$

On the average we would expect an element to be compared with about half the elements in the list before finding its correct place. The total number of comparisons on the average is

$$C = \frac{1 + 2 + \cdots + n - 1}{2} = \frac{n(n - 1)}{4}$$

MERGE

Merging involves combining two or more ordered lists so that the combined list is also in order. It does *not* mean combining the elements and then ordering them. (A merge algorithm is given after the discussion of the two-way merge sort.) Merging is competitive with other internal sorts in terms of speed, but requires twice the amount of storage to merge the numbers from one pass to another. The steps in the two-way merge sorts are:

1. Merge single items into pairs so that the numbers in each pair are in order.

2. Merge pairs of pairs so that the resulting quadruples are in order.

3. Merge pairs of quadruples and continue in this manner until no more merges can occur.

The number of passes in the sort is approximately $\log_2 n$, where n is the number of elements in the list. Figure 7.3 shows the two-way merge sort of a list of 16 numbers. It requires four passes to order the list. Note that storage of the values alternates between the original array and a spare array.

The following MERGE algorithm shows the steps needed to merge two lists, A[1] . . . A[N] and B[1] . . . B[M], to form a third list, C[1] . . . C[N + M]. It assumes that the original lists are arranged in ascending order.

Algorithm MERGE

1. Comments: Merge two lists in ascending order into a single list, also in ascending order.

    Input: A[1] . . . A[N]   A dense list of numbers in ascending order
                  B[1] . . . B[M]   A dense list of numbers in ascending order
    Output: C[1] . . . C[N + M]   A merger of lists A and B in ascending order
    Variables: I, J, K   Index for dense lists A, B, and C

2. Initialize I, J, K to 1:
    I ← 1; J ← 1; K ← 1

3. Repeat the following until the last element of list A or B has been merged into list C:
    If A[I] < B[J] then
        Copy A[I] into C[K]:
            C[K] ← A[I]

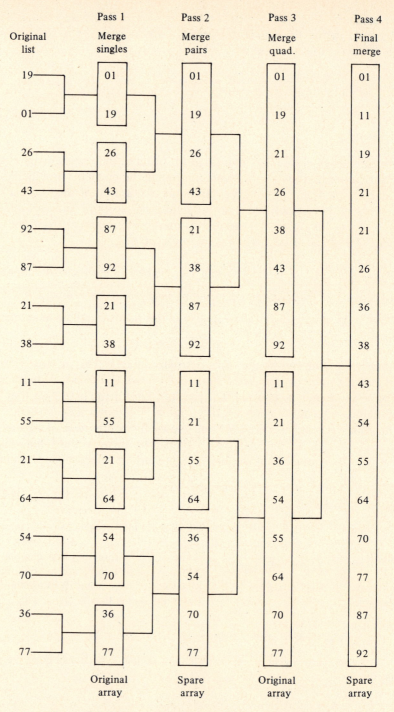

Figure 7.3 Merge sort of 16 numbers in four passes

Increment indexes to work with next A and C:
   I ← I + 1; K ← K + 1
Else
   Copy B[J] into C[K]:
      C[K] ← B[J]
   Increment indexes to work with next B and C:
      J ← J + 1; K ← K + 1
   If I > N or J > M, proceed to step 4.

4. If all of list A has been merged, then
   Copy remainder of list B into C:
      C[K] . . . C[N + M] ← B[J] . . . B[M]
Else
   Copy remainder of list A into C:
      C[K] . . . C[N + M] ← A[I] . . . A[N]

5. End MERGE

HEAP SORT

The heap sort imposes a tree structure on the values as it orders them and
makes use of the dense list numbering of tree nodes depicted in Figure 5.11.
Initially it builds a heap out of the numbers so that the largest value appears
at the root of the tree. A *heap* is an ordering such that every node is larger
than its immediate successors. In particular this means that the root is the
largest of all the numbers.

Figure 7.4 gives an example of a heap. The small numbers outside each
node show the node's index in the dense-list storage of a tree. Remember
that node i's successors are numbered $2i$ and $2i + 1$. The nodes in the tree
are not completely ordered yet. That is, there is no scan order that will produce
the numbers in ascending or descending order. We must work with the heap
until all numbers are ordered.

After the heap is built, we want to output or set aside the largest value
and work with the remaining nodes on the tree. We find the second-largest
value and set it aside. We repeat our process, each time decreasing the size
of the tree and reducing our sort effort. Our strategy is to set aside the largest
value as it is found, so that all the nodes can be output at the end when they

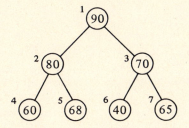

Figure 7.4 A sample heap

are in order. We choose the last node in the tree as the place to set aside the largest, and temporarily remove the last node to an extra location until we can discover where to place it in the heap. In Figure 7.4, this means that the 65 in position 7 is removed and replaced with 90. This leaves the root position empty and 65 waiting in EXTRA for a place to go.

Figure 7.5(a) uses a dashed line to show that the last node has been pruned from the tree. There is an empty position to fill and node 65 is temporarily outside the tree. The choices for the new root are either node 2, 3, or the EXTRA value. Since the 80 at node 2 is largest, we move it up to the root position. To fill position 2, we examine its successors, nodes 4 and 5, and the EXTRA node. Since 68 is largest, it moves up to position 2. There are no successors of 68, so the EXTRA value 65 takes its place. Figure 7.5(b) illustrates the tree with the second-largest node at the root.

Once again we place the root in the last node. Since our tree now has only six nodes, we place 80 in position 6 and set 40 aside. This leaves our tree as shown in Figure 7.5(c). We continue in this manner to fill the root position and place the last node in its correct place in the heap. Finally we end up with the tree shown in Figure 7.5(d). The nodes are in ascending order from node 1 up through node 7. Alternatively, beginning at node 7 and working backward, we have a descending order.

The following HEAPSORT algorithm describes the steps we have performed in the example. It can work with a binary tree of any size.

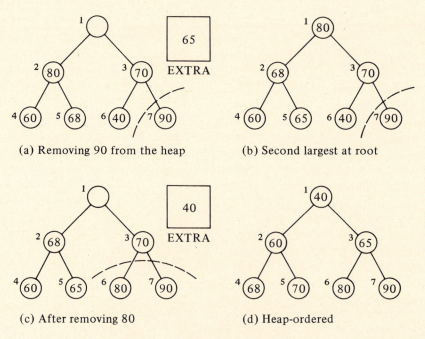

(a) Removing 90 from the heap

(b) Second largest at root

(c) After removing 80

(d) Heap-ordered

Figure 7.5 Steps in a heap sort

Algorithm   HEAPSORT

1. Comments: Order a dense list in ascending order.
    Input: D[1] . . . D[N]   A dense list to be ordered
                  N   The number of elements in the list
                  LEV   The number of levels in the tree
                  R   Index to rightmost node in tree
    Output: D[1] . . . D[N]   The list in ascending order; D[1] is the smallest;
                                     D[N] the largest
    Variables: I   Index to D
                  EXTRA   An extra location outside the tree

2. Convert the list into a heap:
    Repeat the following for all levels in the tree beginning with level LEV − 1 and
    proceeding up to level 1, the root (level LEV − 1 begins with index N/2 of
    array D):
        For each node D[I] in the level, compare D[I] with its immediate successors
        D[2I] and D[2I + 1] (if they are part of the tree):
            Set D[I] ← largest of D[I], D[2I], and D[2I + 1]; D[2I] and D[2I + 1]
            contain the other two values
            If D[I]'s value has changed, go down the branch D[I] came from and verify
            that the heap relationships still hold.

3. Order the heap:
    Repeat the following for each node R, with R beginning at N and proceeding
    down to 2:
        Set node at position R in position 1; set root node in position R and in EXTRA:
            EXTRA ← D[1]
            D[1] ← D[R]
            D[R] ← EXTRA
        Make a heap with nodes 1 through R − 1.

4. Output the tree:
    Output D[1], D[2], . . . , D[N]

5. End HEAPSORT

The number of comparisons required is on the order of $n \log_2 n$, which
approaches the theoretical minimum number of comparisons. The actual number of comparisons and the number of times that data is moved depends to
a certain extent on the original order of values. If, for example, the values
were in reverse order, then the creation of the heap would require no moving
at all. This is probably contrary to our intuition, which would tell us that if
the data are in reverse order, there should be more work required to put them
right.

Implementing the HEAPSORT algorithm, we see that subscripts of 2I and
2I + 1 appear frequently and we may be tempted to program the subscript
as 2 * I and 2 * I + 1 for the sake of clarity. However, depending on the
language and the compiler we are using, a subscript of 2 * I may cause a

multiply operation, which is one of the more time-consuming operations. A more intelligent compiler may generate a shift left 1 position, which is equivalent to a multiply by 2 on a binary number. If the index is not stored in binary format, then the shift will not work and a multiply is necessary.

A solution to this problem is to introduce a new index J, perform the operation J := 2 * I once and then reference the variables as D[J] and D[J + 1]. Following is an implementation of the HEAPSORT algorithm that attempts to simplify its reference of indexed variables. Note the subtle differences between the English-like algorithm and the Pascal statements. For example, steps 2 and 3 of the algorithm require that the list becomes a heap. A HEAP procedure was therefore written for the program. The program makes use of the HEAP procedure repeatedly during its operation.

Initially, it begins at level LEV − 1 (the middle of the list) and makes the lower part of the tree into a heap. It repeats the heaping operation, including one more node from the list with each call of HEAP. Finally the entire list is a heap. The second half of the program (step 3 of the algorithm) swaps the root and the last node in the tree. The last node is also placed in EXTRA, so that at every empty node I in the tree we can place the largest of D[2I], D[2I + 1], and EXTRA in node I. To understand the algorithm and its implementation, we suggest that you trace through the program with some sample values.

```
program HSORT (INFILE,OUTPUT);
                                (* Perform heapsort on a dense list (array) *)
                                (* of values                                *)
type
     INDEX = 1 . . 100;
var
     INFILE : text;
     D        : array [1. .100] of integer;
     EXTRA : integer;
     N,                         (* Number of elements in list            *)
     I,                         (* Index to element in list              *)
     L,                         (* Left index of sublist to heap         *)
     R        : INDEX;          (* Right index of sublist to heap        *)
procedure HEAP (I : integer);   (* Create a heap from elements I through R *)
var
     J          : integer;
begin                           (* Begin HEAP procedure                  *)
  J        := 2 * I;            (* Locate left successor of I            *)
  EXTRA := D[I];
  while J <= R do
    begin
      if J < R then
        if D[J + 1] > D[J] then (* Find index of largest successor       *)
          J := J + 1;
        if EXTRA >= D[J] then   (* Found place where EXTRA goes          *)
```

```
        begin
          D[I] : = EXTRA;
          J    : = R + 1;            (* To force exit from HEAP procedure    *)
        end
      else
        begin
          D[I] : = D[J];             (* Proceed  down  subtree  of  I's largest *)
                                     (* successor                            *)

          I    : = J;
          J    : = 2 * I;
        end;
      end;
    D[I] : = EXTRA;                  (* Put EXTRA at a terminal node         *)
  end;                               (* End HEAP procedure                   *)

  begin                              (* Begin HSORT program                  *)
    read (INFILE, N);                (* Input number of values               *)
    for I : = 1 to N do
      read (INFILE, D[I]);           (* Input array to order                 *)
    L : = (N div 2) + 1;             (* Locate next-to-last level            *)
    R : = N;                         (* Start at rightmost node in tree      *)
    while L > 1 do                   (* Create heap out of tree elements L through *)
                                     (* R                                    *)

      begin
        L : = L - 1;
        HEAP(L);
      end;
    while R > 1 do
      begin
        EXTRA : = D[1];              (* Set aside root of tree               *)
        D[1] : = D[R];
        D[R] : = EXTRA;              (* Last node is former root             *)
        R    : = R - 1;             (* Prune last node from tree            *)
        HEAP(1);                     (* Heap nodes 1 through R               *)
      end;
    for I : = 1 to N do              (* Output values in ascending order     *)
      writeln (OUTPUT, D[I]);
  end.                               (* End of HSORT program                 *)
```

QUICKERSORT

In the bubble-sort algorithm, the distance that an element moves after a comparison is either 0 or 1 position. Thus if an item is at the wrong end of the list, it will take many comparisons and moves to locate it in the correct place. In the selection sort one item may move quite a few spaces, but at the expense of possibly relocating another element at the wrong end of the list. The purpose of quickersort is to cover a lot of distance in one move in the correct direction.

Basically quickersort selects the middle element of the list and divides the remainder of the list into two parts—one part containing all values less than the middle element, and the other part containing all values greater than the middle element. These two parts (sublists of the original list) may or may not be the same size, and values equal to the middle element may go in either part. The next step is to store one of the sublists on a stack and then work on the other sublist in the same manner. Thus we are constantly working on smaller and smaller sublists. When left with a sublist of length two (anything less than ten is just as efficient), order the elements and process one of the sublists from the stack. When the stack is empty, the entire list is in order. To ensure that this process is not magical, let us study the following numerical example. Then we will go into the details of the algorithm.

Suppose that we are given the following 13 values:

$$20 \quad 73 \quad 42 \quad 11 \quad 80 \quad 39 \quad ⑦② \quad 30 \quad 100 \quad 46 \quad 88 \quad 32 \quad 21$$

The "middle" of the list is the seventh number, 72, which we circle. Next we begin at the left end looking for a value greater than 72 and at the right end looking for a value less than 72. From either end we proceed sequentially. If the numbers are stored in a dense list, the left-hand search begins with index 1 and proceeds to 2, 3, and so on. The right-end search begins with index equal to 13 and decreases the index by 1 until a value less than 72 is found.

$$20 \quad 73 \quad 42 \quad 11 \quad 80 \quad 39 \quad ⑦② \quad 30 \quad 100 \quad 46 \quad 88 \quad 32 \quad 21$$
$$\quad \uparrow \uparrow$$

The arrows point to numbers that satisfy this rule. We now exchange these values and continue our search for two more numbers.

$$20 \quad 21 \quad 42 \quad 11 \quad 80 \quad 39 \quad ⑦② \quad 30 \quad 100 \quad 46 \quad 88 \quad 32 \quad 73$$
$$\quad \uparrow \uparrow$$

We swap these and continue, but do not consider 72.

$$20 \quad 21 \quad 42 \quad 11 \quad 32 \quad 39 \quad ⑦② \quad 30 \quad 100 \quad 46 \quad 88 \quad 80 \quad 73$$
$$\quad \uparrow \quad \uparrow$$

After swapping 100 and 46, we realize that we cannot continue and have divided our list into two parts. We must move our "middle" element into its correct position between 46 and 100.

$$\underbrace{20 \quad 21 \quad 42 \quad 11 \quad 32 \quad 39 \quad 30 \quad 46}_{\text{sublist}_1} \quad ⑦② \quad \underbrace{100 \quad 88 \quad 80 \quad 73}_{\text{sublist}_2}$$

Since sublist₁ is longer than sublist₂, we will apply the same algorithm to it, leaving 72 in its correct position and temporarily leaving sublist₂ as is. Picking 32 as the "middle" of sublist₁, we are able to form two new sublists.

$$20 \quad 21 \quad 30 \quad 11 \quad \underbrace{\textcircled{32}} \quad \underbrace{39 \quad 42 \quad 46} \quad \textcircled{72} \quad \underbrace{100 \quad 88 \quad 80 \quad 73}$$
$$\underbrace{}_{\text{sublist}_3} \qquad \underbrace{}_{\text{sublist}_4} \qquad \underbrace{}_{\text{sublist}_2}$$

We could continue working on each sublist, but since there are so few numbers in each, we might as well point out a few things. First of all, 32 and 72 are in their correct positions, according to the final ordering. All sublists are in their correct positions, relatively speaking. That is, all numbers in sublist₃ are less than 32. All numbers in sublist₄ are between 32 and 72. And all numbers in sublist₂ are greater than 72. Thus, if we order each sublist, we will obtain the desired result.

$$11 \quad 20 \quad 21 \quad 30 \quad 32 \quad 39 \quad 42 \quad 46 \quad 72 \quad 73 \quad 80 \quad 88 \quad 100$$

Perhaps one of the most subtle tricks in the quickersort algorithm involves handling the middle element. In the example, 72 started out in the seventh position and was inserted in the ninth position between 46 and 100. An insertion of this type means moving elements and moving requires time. To eliminate moving a lot of elements, temporarily remove the middle element, place it in an area outside the list, fill its spot with the first element in the list, and begin processing with the second element. After splitting the list into two parts, place the middle element back into the list and put an appropriate value in the first position. Our example would have started in the following way: The first element replaces 72 and 72 is set apart from the list.

$$\boxed{20} \quad 73 \quad 42 \quad 11 \quad 80 \quad 39 \quad 20 \quad 30 \quad 100 \quad 46 \quad 88 \quad 32 \quad 21 \quad \textcircled{72}$$

After making all swaps, the list is now in the following arrangement.

$$\boxed{20} \quad 21 \quad 42 \quad 11 \quad 32 \quad 39 \quad 20 \quad 30 \quad 46 \quad 100 \quad 88 \quad 80 \quad 73 \quad \textcircled{72}$$
$$\uparrow \qquad \uparrow$$

Now 46 is at the end of the left sublist. Since 46 is less than 72, we move 46 to position 1 and let 72 occupy the place of 46. We still have two sublists—one with numbers less than 72 and one with numbers greater than 72.

$$\underbrace{46 \quad 21 \quad 42 \quad 11 \quad 32 \quad 39 \quad 20 \quad 30}_{\text{sublist}_1} \quad 72 \quad \underbrace{100 \quad 88 \quad 80 \quad 73}_{\text{sublist}_2}$$

If the searches ever meet at one end, then we can shorten our sublist by only one element. If both searches meet at the right end, then all values in the list are less than the middle value. On the other hand, if both searches meet at the left end, then all values in the list are greater than the middle value. Thus the middle value gets positioned at one end of the list and the sort can continue on the remaining values.

If we find a number that belongs in the right sublist, but do not find a number to swap with it, we place the middle value in front of that number, after removing the number previously stored there and placing it in position 1. For example, suppose that we begin with the following list.

45 73 12 80 100 62 48 60 (46) 71 89 33 92 87 36 61 50

We remove 46, the middle element, and put 45 in its place. The first position is marked to show that it is not under consideration.

```
[45]  73  12  80  100  62  48  60  45  71  89  33  92  87  36  61  50  (46)
      ↑                                                     ↑
```

After swapping 73 and 36, we go on to swap 80 and 33, and then 100 and 45. Now we discover that 62 should go on the right sublist, but that there is nothing to swap with it.

```
[45]  36  12  33  45  62  48  60  100  71  89  80  92  87  73  61  50  (46)
                      ↑
```

Therefore 62 becomes the beginning of the right sublist, if we put 46 in the position occupied by 45 and place 45 in the unused first position.

```
45 36 12 33    (46)    62 48 60 100 71 89 80 92 87 73 61 50
 sublist_1                         sublist_2
```

We have thus separated the original list into two sublists and can continue with the algorithm. Since we can work with only one sublist at a time, we save one sublist's left and right pointers and work with the other sublist. In this example, we stack the left sublist pointers (1 and 4) and work with the right sublist. Pointers for sublist$_2$ are 6 and 17 for the left and right ends, respectively. After processing sublist$_2$, using 89 as the middle, we have the following.

```
                                                                  stack
45 36 12 33  (46)  73 48 60 50 71 62 80 61 87  (89)  92 100
 sublist_1              sublist_3                   sublist_4     [1,4]
```

As before, we stack the left sublist pointers (6 and 14) and work with the right sublist. Since there are only two elements in sublist₄, we order them and remove a sublist from the stack.

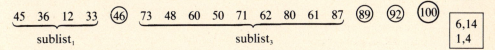

stack

| 6,14 |
| 1,4 |

After processing sublist₃ we now find that all circled elements are in their correct locations and are no longer considered in the sorting. We stack the pointers for sublist₅ and are ready to work on sublist₆.

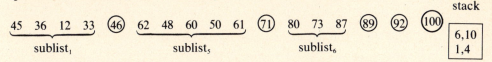

stack

| 6,10 |
| 1,4 |

By now you should be able to continue this example on your own and verify that the quickersort algorithm does produce an ordered list.

Algorithm   QSORT

1. **Comments:** Sort an array of values using the quickersort technique. Place the values in ascending order. QSORT uses recursion to work on smaller and smaller sublists until the entire array is ordered. Initially QSORT is called to work on the entire array.

   Input: L   Left index of sublist
                R   Right index of sublist

   Output: X[L] . . . X[R]   Sublist in order

   Variables: X   Array to be ordered
                M   Middle element
                IL   Index for search from left side
                IR   Index for search from right side

2. If list contains three items or less (R − L < 3), then
   Order the values
   EXIT QSORT

3. Find middle element of array X; use integer division and discard any remainder:
   M ← X[(L+R)/2]

4. Place first element in middle position:
   X[(L+R)/2] ← X[L]

5. Initialize search indexes:
   IL ← L + 1; IR ← R

6. Repeat the following:
   (a) Search from left so that IL is set to an index of a value greater than middle value; that is, X[IL] > M.
   (b) Search from right so that IR is set to an index of a value less than middle value; that is, X[IR] < M.

(c) If IL < IR (left index is still to left of the right index), then,
    Swap X[IL] and X[IR]
    Increase IL by 1:
        IL ← IL + 1
    Decrease IR by 1:
        IR ← IR − 1

(d) If IL ≥ IR (search indexes have crossed), then
    Proceed to step 7

7. Place middle element back in list:
    X[L] ← M

8. If L < IR, then
    Call QSORT with left index L and right index IR

9. If IL < R, then
    Call QSORT with left index IL and right index R

10. End QSORT

The expected number of comparisons C for quickersort is:

$$C \approx 1.4\ n\ \lceil\ \log_2\ n$$

where there are n elements in the list and $\lceil$ means "increase the value up to the nearest integer." It has been determined empirically that quickersort is the fastest internal comparative sorting technique on most machines.

SUMMARY

To compare sorting algorithms, we must consider many factors: the volume of data to handle, the amount of storage required, and the desired speed. Is it more important that the algorithm be executed quickly, or that it be implemented in a short amount of time? When implementing an algorithm, we must evaluate the number of times the sort loop (or loops) is executed. It is also important to know what happens inside that loop, including the number of comparisons, the amount of data that gets moved, and the amount of overhead involved in computing indexes and locating data. The cost of recursive routines or subroutine calls are also important considerations.

An interesting way to evaluate sorts is to select several sort algorithms, implement them, and test them against different types and amounts of data. All the algorithms to be evaluated can be programmed and run against the same sets of data. The experiments should include both large and small samples of data, perhaps with both numeric and alphanumeric keys. The discussion of the number of comparisons performed by various algorithms in this chapter suggests that some sample data be arranged to exercise the sort and produce either the minimum or maximum amount of comparisons. For some sorts, this may mean that the data must be presorted. For others, it means that it is in reverse order. For example, heapsort is expected to perform at its worst if the data are in the opposite order from that desired. To generate sample data, we can use the system's random-number generator or write our own data generator. Other variations for data include lists with many duplicate values, or lists interleaving ascending and descending values. For example, the first, third, fifth, and so on values could be ascending, while the second, fourth, and every alternate value are descending.

To compare the sorts, we can time the execution or measure the number of instructions executed. Common measures of time are CPU time (time that the CPU is busy doing work) and elapsed time (total job time from start to finish, including any wait time).

After making the runs on our sample data, we must analyze the results. Unfortunately, making comparisons will be difficult when some sorts perform well on certain data but less well on other data. This only emphasizes the fact that we need to know something about the algorithms and something about the data before we can say that one sort is better than another.

EXERCISES

1. Write a sort algorithm for one (or more) of the sorting techniques in the section.

2. Write a program for one (or more) of the sorting techniques in the section. Run it with different sets of data and compute the average number of comparisons. Compare your program's average with the theoretical average.

3. Alter the MERGE algorithm so that it can be used in a two-way merge sort.

4. What are some of the considerations involved when sorting names or other alphanumeric data? Write a program to sort a list of names. Use any sort algorithm desired.

5. Complete the execution of the HEAPSORT algorithm on the tree in Figure 7.5.

6. Compare HEAPSORT and QSORT when the original list (7, 5, 4, 3, 2, 1) is in reverse order.

7. Plot the number of comparisons versus the size of the list for the algorithms for bubble sort and quickersort. Vary the size of the list from 100 to 5000 items. Also plot the theoretical minimum number of comparisons.

8. Explain how the RADIX SORT can be used to alphabetize a set of names, each a maximum of 10 characters.

9. Design an experiment to compare sort algorithms.

10. Implement the HEAPSORT algorithm. Does your implementation differ from that in the text?

11. Implement QSORT, the quickersort algorithm.

12. Rewrite the QSORT algorithm so that it is not recursive.

13. Implement the algorithm in Exercise 12.

14. Investigate the methods of timing programs that your computer system offers.

15. Modify the HEAPSORT algorithm so that it operates on a linked-list representation of a binary tree.

7.3
External Sorting

Whenever the volume of data to be sorted is too great for all of it to fit in memory at one time, we must utilize some external device such as disk or tape. We store sorted portions of the data on the external device and work with as much as possible in internal memory. Most external sorts are actually a sort/merge, as depicted in Figure 7.6. Initially, as much of the input file as possible is read into an input buffer area. The input is sorted and written to a work file. This allows room for more of the input data to be read, sorted, and saved on the external device. Then, when all portions of the input have

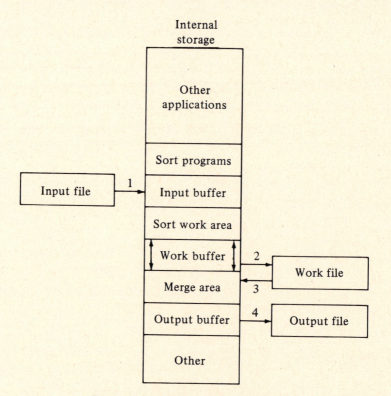

Figure 7.6 Diagram of a sort/merge

been processed, the merge phase begins. It merges together the sorted parts, repeatedly if necessary, until the entire collection is merged into the output file. Thus, the transfer shown in Figure 7.6 between the work buffer, work file, and merge area (arrows 2 and 3) may occur many times.

Many of the external sorting algorithms use either disk or tape for their work files. Some are restricted to disk devices because they need to retrieve the data in a random order. Tape drives provide only sequential access. Initial input to or final output from the sort can be on tape, on disk, or on any suitable medium. The work files can be allocated to different disk drives or to the same one. If the computer system has multiple-disk drives and it is possible to overlap input-output time by using them simultaneously during a sort, then we can take advantage of the various drives. If, however, our work file area is on a single disk, then the sort may cause much arm movement on the disk as different areas are accessed.

Elements of the sorting phase include the internal sorting algorithm, the storage allocation, and the dispersion algorithm used to write or disperse the strings to the external devices. In sorting terminology, a *string* is a collection of data that is ordered. Another term for a string is a *run*. Making efficient use of storage is also an important part of the sort or merge phase. The sort or merge program and the input and output areas, together with the sort or merge work area, are all competing for storage. Figure 7.6 shows a separate sort and merge work area. Since the sort operation is first, it can share the same work area as the merge. Also, the input buffer and output buffer can be the same, since input occurs in the sort phase and output of the ordered file occurs during the merge phase.

Characteristics important to the merge phase are the number of strings to be merged, the string length, the number of drives used to disperse the strings, and the blocking characteristics of the work files. Merge *order* is the number of strings that can be compared and merged into one string during a merge operation. For example, the binary radix sort has a merge order of 2. At any time two groups of data are combined into one. If we have 30 strings and a merge order of 2, then we must make five passes over the entire set of data to reduce the 30 strings down to one. The first pass combines the 30 strings, two at a time, down to 15 strings. The next pass merges the 15 strings to 8. Then the last three passes convert 8 strings to 4, 4 strings to 2, and 2 strings to 1. In general, the two-way merge (merge order 2) takes p passes over the data, where

$$p = \lceil \log_2 n$$

and n is the number of strings. If the merge order were 4, then we would need three passes over the data. For a merge order of m and n strings to merge, the number of passes required to produce one string is

$$p = \lceil \log_m n$$

The size of the sort work area determines the number of strings produced, which in turn directly affects the number of merge passes. The more strings there are, the more passes over the data the merge phase must make. Producing fewer strings requires a large sort area, but that may decrease the amount of storage for input-output (I/O) buffers. A smaller I/O buffer may cause more I/O operations. The number of records that fit in the sort area depends on the size of the area and the record length. The sort program can define the size of its sort area, but has no control over the record length. Because external sorts generate I/O operations to their work files and because the time for an I/O operation is large compared with the time it takes to execute a compare or move instruction, the analysis of external sorts is different from that of internal sorts. Even though we desire a sort that executes in the least amount of time, for external sorts we must include I/O time as well as CPU time. The efficient external sort algorithm must consider blocking of data to reduce the number of I/O operations. Also, it must balance sort time and merge time to minimize the total time for a complete sort. A fast sort phase may result in partially ordered data that requires many merge passes. On the other hand, having few merge passes may require much time or storage for sorting the records.

REPLACEMENT SELECTION

The replacement selection sorting technique is very commonly used. It is similar to the heap-sort methodology. As the largest (or smallest) value is selected, it can be transferred to an output buffer area and replaced by another record. A sorting technique then determines the largest record in the sort area. If this record's key is not greater than the key of the record already output, it can be added to the sort string. We can continue adding a record to the sort area, locating the largest key, and outputting it so long as the keys remain in descending order. When the sort area contains records with keys larger than the last key output, the first string ends and we start generating a new string. If the sort area can hold n records, then the average string length is $2n$.

In a replacement selection sort the number of records being sorted remains the same until the end of data is reached. Data are transferred to a work file area whenever a buffer fills up. A string may thus span several blocks of data on the external device, and the sort must keep track of the location of the strings.

Internally the sort work area contains the records being ordered. To be efficient it may construct a sort tree containing pointers to those records. The sort therefore uses the pointers to locate the keys for its ordering, but it moves only the pointers. Whenever the largest key is determined, the pointer in the sort tree locates the record to be transferred to the output string.

Replacement selection produces strings of various lengths that can be merged optimally into one ordered string by forming a minimal merge tree. Such minimal trees are called *Huffman trees* in coding theory. They represent the fastest decoding trees for coded messages.

OPTIMAL MERGE TREE

Suppose that we use a replacement selection algorithm to produce strings of lengths 28, 25, 13, 10, 8, 7, 6, and 3. Of the eight strings, we arbitrarily select three of them to be merged into an empty area. The following question arises: Which three strings do we merge first?

Suppose that we start with the largest strings first and merge them together, thereby creating an ordered list of length 28 + 25 + 13 = 66. The strings to be merged next are of lengths 66, 10, 8, 7, 6, and 3. If we continue to merge the three largest strings first, we produce 84, 7, 6, and 3; then 97 and 3; and finally 100.

We can draw a tree representing the steps in merge sorting, using three strings at a time. This tree contains terminal nodes that represent the lengths of the strings. Internal nodes represent merged strings, and their value is the sum of the values of their successor nodes. That is, their value is the length of the merged strings. See Figure 7.6(a) for the merge tree created by merging, as we have suggested above. The root is at level zero to make our calculations easier.

Suppose that we perform the merge as shown by the tree in Figure 7.7(b). The first strings merged are the shortest strings of length 6 and 3. They form a string of length 9 that is merged with the next smallest strings of lengths 8 and 7. The resulting string of length 24 is merged with strings of length 13 and 10. We finally obtain the ordered string of length 100, as we did before.

The merge tree of Figure 7.7(b) is called a *minimal,* or *optimal, merge tree* because fewer move operations are needed when merging in the order shown. How can we determine this order, and how can we build optimal merge trees?

To determine the number of moves, we notice that every level in the merge tree represents a move (merge). To move the elements of an initial string to the final string, we must merge the terminal nodes repeatedly until we merge into the root node. Hence to move an element from an initial string, we must elevate it from its level in the merge tree to the root level.

A terminal node at level L must be moved L times to reach the root level. If a terminal node has a value of v_i, Lv_i is the number of moves needed to copy all v_i elements from the initial string to the final string.

Let L_i be the level number of terminal node i and v_i be the value (length) of terminal node i. The value of a tree with r levels is

$$\text{Value} = \sum_{i=1}^{r} L_i v_i$$

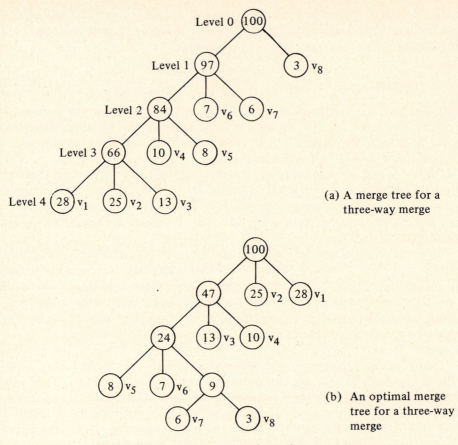

(a) A merge tree for a three-way merge

(b) An optimal merge tree for a three-way merge

Figure 7.7 Two merge trees for a three-way merge

A minimal or optimal merge tree is a tree constructed with terminal nodes (representing strings of length v_i) arranged so that the value of the merge tree is as small as possible.

The values of the trees of Figure 7.7 are shown by the computations in Figure 7.8. In each case the value of nodes at level n are summed and multiplied by the level number to give the total value at that level. The level products are summed to give the value of the tree.

The formula for merge tree value gives us a means of checking the value of a tree to determine whether it is minimal. The second problem is to determine how we construct minimal trees.

A two-way merge produces a binary tree; a three-way merge produces a ternary tree. In general, a c-nary tree represents a c-way merge. A minimal-merge c-nary tree is constructed by grouping together the c nodes of smallest value and calling these the *filial set*. The c smallest values are summed, and

(a) Tree of Figure 7.7(a)

Level number	Nodes	Values	Product
1	v_8	3	3
2	v_6, v_7	6 + 7	26
3	v_4, v_5	10 + 8	54
4	v_1, v_2, v_3	28 + 25 + 13	264
		Value =	347

(b) Tree of Figure 7.7(b)

Level number	Nodes	Values	Product
1	v_1, v_2	28 + 25	53
2	v_3, v_4	13 + 10	46
3	v_5, v_6	8 + 7	45
4	v_7, v_8	6 + 3	36
		Value =	180

Figure 7.8 Computation of values for two merge trees

the sum becomes the value of the father node of the filial set. Again we select the c smallest nodes from all nodes remaining and call this the filial set. We form the father of this filial set and assign it the value equal to the sum of the filial set values. This procedure is repeated until only one node remains.

Often the number of terminal nodes does not equal a multiple of c. When this is the case, we must select the size of the first filial set as follows. For a tree with r levels, if $(r - 1)$ modulo $(c - 1) \equiv 0$, the first filial set contains c nodes, otherwise the size of the first filial set is

$$1 + [(r - 1) \text{ modulo } (c - 1)]$$

An example of constructing the optimal tree of Figure 7.7(b) is given in Figure 7.9. Since $r = 4$ and $c = 3$, the first filial set size is

$$1 + [(4 - 1) \text{ modulo } (3 - 1)] = 2$$

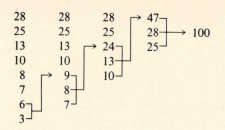

Figure 7.9 Derivation of minimal merge tree

TAG SORT (Key Sort)

In a *tag sort* (also known as a *key sort*) the unordered file has its keys (or tags) copied onto an extra file, together with pointers to the original data records. The sort/merge then works with the tag file, placing it in order. The ordered tag file then shows how to read and write the records in correct order. This method may appear to be reasonable, but it takes much I/O time to write the final ordered file.

The unfortunate part of tag sorting is the rearrangement of the master file records necessary after the sort. Each record must subsequently be moved to its proper place in order.

It has been theoretically shown that the rearrangement overhead induces as much overhead as sorting without splitting off the keys first. For a file of N records, B records per file block, and a main memory buffer of M records, we must perform about

$$\frac{N \log_2 B}{M}$$

block reading operations to rearrange the records after the key sort.

PEER SORT

The peer-sort technique is designed for files that are partially ordered, which is often the case for disk files. The sort phase records the location of the strings and the keys they contain. If possible, strings can then be concatenated to form longer strings, before the merging takes place. Strings are subdivided into string *segments,* where a segment is that portion of the string that will fit on a track. An index records the location of the string segment, its lowest key value, and its highest key value. The peer-sort algorithm operates on the string index in the following way.

1. Sort the strings from low to high

2. Start a new string with the string containing the lowest key value (the first index entry)

3. Repeat until all strings have been processed:
    If the next index can extend the current new string,
        Concatenate the index entry to the string
    Else
        Start a new string

Next, a merge phase consolidates the strings. With a merge order of m, the strings are divided into groups containing m strings. The first group has $m - 1$ strings. Each group is merged to one string. If one more merge would complete the sort/merge, it is performed. Otherwise, the sort repeats its string definition and merge phase as many times as necessary.

BALANCED MERGE

The balanced-merge technique was initially designed when tape drives were the primary external storage media for external sorts. It used as many tape drives as possible, but could be adapted to multiple disk drives or multiple sort work files on disk. Before the merge begins, its sort phase produces an equal number of ordered strings on half the external work areas (half the tape drives or half the disk work areas). The strings may be of the same length or of different lengths. To describe the balanced merge, we present an example in which there are four tape units and six sorted substrings. We have kept our numbers small so that we can explain each merge pass. The substrings are written on two of the tapes. Thus at the end of the sort phase of the balanced merge sort, the four tapes are as follows: tape 1 is empty, tape 2 is empty, tape 3 has three substrings, tape 4 has three substrings. The tapes are now balanced, either containing ordered substrings or being empty.

In the next phase the first substrings from tape 3 and tape 4 are merged and written onto tape 1, leaving two substrings on both tape 3 and tape 4. Next the second substrings from tapes 3 and 4 are merged onto tape 2, leaving one substring on each. The remaining substrings are finally merged onto tape 1, leaving tapes 3 and 4 empty. At the end of this phase, tape 1 has two (larger) substrings, tape 2 has one (larger) substring, and tapes 3 and 4 are empty. Merging continues until one tape contains the entire (ordered) string (see Figure 7.10 on pp. 168–169).

Repeated passes and merges reduce the string to one sorted string in $\log_T s$ phases, where s is the number of initial strings, t is the number of tape units, and $T = t/2$. This method is similar to the internal merge sort, except that here half the tapes are merged at once.

CASCADE

The cascade merge sort operates on all tapes at once. It first distributes the initial sorted strings on all but one of the tapes and then merges them onto the remaining tape. The merge stops when one of the tapes runs out of data.

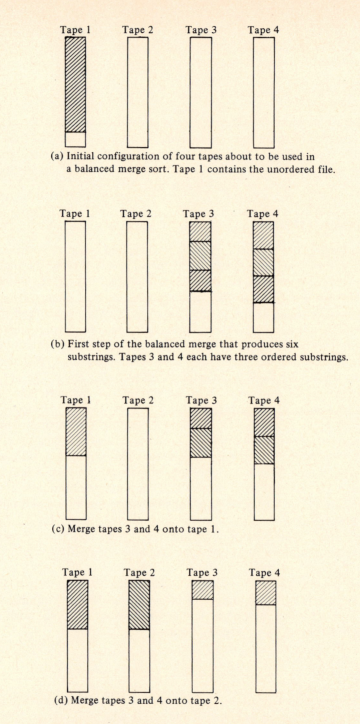

(a) Initial configuration of four tapes about to be used in
   a balanced merge sort. Tape 1 contains the unordered file.

(b) First step of the balanced merge that produces six
   substrings. Tapes 3 and 4 each have three ordered substrings.

(c) Merge tapes 3 and 4 onto tape 1.

(d) Merge tapes 3 and 4 onto tape 2.

Figure 7.10. Configuration of a balanced merge sort

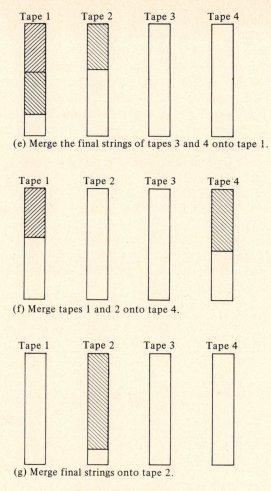

(e) Merge the final strings of tapes 3 and 4 onto tape 1.

(f) Merge tapes 1 and 2 onto tape 4.

(g) Merge final strings onto tape 2.

Figure 7.10 (continued)

Assuming that there are t tapes, we find that a subsequent pass begins by merging $t - 2$ of the unmerged tapes onto an empty tape. Repeatedly $t - 3$, $t - 4$, and so on tapes are merged until none remain. The final phase is to copy all the merged tapes onto one. In most cases, the cascade merge and balanced merge are not as efficient as the polyphase and oscillating merge, which we discuss next.

POLYPHASE

For installations with six or fewer tape drives, the polyphase sort is best.[1] The initial strings are distributed on $t - 1$ tapes in amounts corresponding

[1] D. L. Shell, "Optimizing the Polyphase Sort," *Comm. ACM* 14, no. 11 (November 1971):713–719.

to a generalized Fibonacci sequence (see Figure 7.11). Like the cascade merge, all $t - 1$ tapes are merged onto the remaining tape until one is emptied. The empty tape is used as the output or merge tape on the next pass. Unlike the cascade merge, each pass uses $t - 1$ tapes.

The number of strings to be merged is computed by a generalized Fibonacci sequence, where each number in the sequence is the sum of the $t - 1$ numbers preceding it. For example, if $t = 4$, the Fibonacci sequence for total number of initial strings is 1, 3, 5, 9, 17, 31, ... It is not often that unordered strings come in Fibonacci sequence sizes, so Shell has offered a near optimum algorithm for the initial tape assignment. The optimum assignment algorithm is very complex and offers only a 3% improvement over the *horizontal distribution* method given by Shell.

A polyphase merge of nine numbers with four tapes appears in Figure 7.12 (pp. 171–172). Pass 0 shows the data in its original order on tape 1. Pass 1 writes the initial strings of length $a_3^{(1)} = 2$, $a_3^{(2)} = 3$, and $a_3^{(3)} = 4$ (these are from the table in Figure 7.11). Pass 2 merges the data to tape 1 until tape 2 is empty. Pass 3 merges the data to tape 2 until tape 3 is depleted, and pass 4 shows the results of the final merge.

k	Total number of strings at start of merge $a_k^{(0)}$	Tape 1 $a_k^{(1)}$	Tape 2 $a_k^{(2)}$	Tape 3 $a_k^{(3)}$
1	3	1	1	1
2	5	1	2	2
3	9	2	3	4
4	17	4	6	7
5	31	7	11	13
6	57	13	20	24
7	105	24	37	44

Figure 7.11 *Polyphase table for* $t = 4$. *The general form of the Fibonacci sequence is*

$$a_k^{(i)} = \sum_{\tau=1}^{t-1} a_{k-\tau}^{(i)}; \quad a_k^{(1)} = 1, \quad i = 2, 3, \ldots ,$$

$$k = t, t + 1, \ldots , \quad a_1^{(0)} = t - 1$$

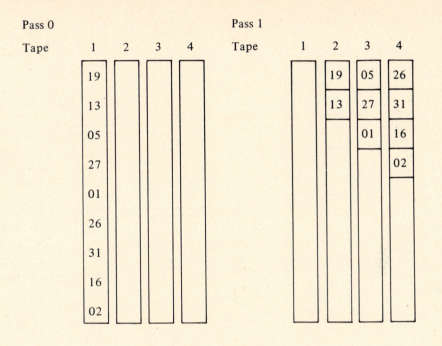

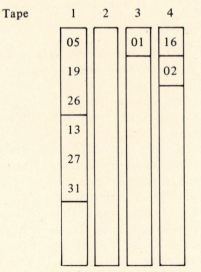

Figure 7.12 Polyphase merge of nine numbers with four tapes

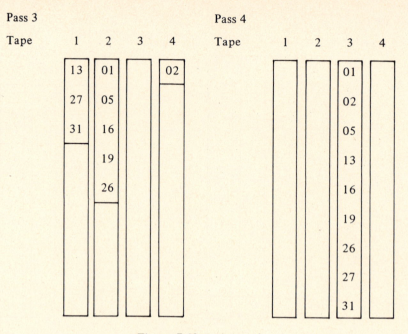

Figure 7.12 (continued)

OSCILLATING SORT

The oscillating sort is applicable *only* to tapes that can be read in either the forward or reverse direction. One string is initially written on all but one of the tape units and then merged to the empty tape. More of the string is then written to the tapes and merged. The term oscillating sort derives from the nature of the switching back and forth from internal to external sorts.

Figure 7.13 illustrates an example of an ascending-order oscillating sort applied to 12 numbers and 5 tapes. The unordered list is initially on tape 1. In the first pass, numbers are written onto all but one of the remaining tapes. The tapes will be read in reverse order instead of being rewound, as shown by the dots in Figure 7.13. The dots indicate the position of the read/write head of each tape drive. During pass 2 the sorted numbers are merged to tape 2, but note that the numbers are merged in descending order. Pass 3 distributes more values to tapes 2, 4, and 5, and pass 4 merges them to tape 3. The sorting and merging continues until all values are merged into ascending order in pass 9. For more detailed information on this and other external sorting methods, refer to Lorin.

Pass 0 Original data

1	2	3	4	5
29	•	•	•	•
23				
15				
27				
01				
26				
31				
16				
02				
46				
10				
22				

Pass 1 Sort

1	2	3	4	5
27	•	(29)	(23)	(15)
01		•	•	•
26				
31				
16				
02				
46				
10				
22				

Pass 2 Merge

1	2	3	4	5
27	29	•	•	•
01	23			
26	15			
31	•			
16				
02				
46				
10				
22				

Pass 3 Sort

1	2	3	4	5
31	29	•	(01)	(26)
16	23		•	•
02	15			
46	(27)			
10	•			
22				

Pass 4 Merge

1	2	3	4	5
31	29	27	•	•
16	23	26		
02	15	01		
46	•	•		
10				
22				

Pass 5 Sort

1	2	3	4	5
46	29	27	•	(02)
10	23	26		•
22	15	01		
	(31)	(16)		
	•	•		

Pass 6 Merge

1	2	3	4	5
46	29	27	31	•
10	23	26	16	
22	15	01	02	
	•	•	•	

Pass 7 Sort

1	2	3	4	5
•	29	27	31	•
	23	26	16	
	15	01	02	
46	(46)	(10)	(22)	
	•	•	•	

Pass 8 Merge

1	2	3	4	5
•	29	27	31	46
	23	26	16	22
	15	01	02	10
	•	•	•	•

Pass 9 Merge

1	2	3	4	5
01	•	•	•	•
02				
10				
15				
16				
22				
23				
27				
29				
31				
46				
•				

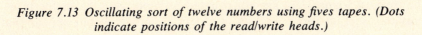

Figure 7.13 Oscillating sort of twelve numbers using fives tapes. (Dots indicate positions of the read/write heads.)

CONSIDERATIONS FOR DISK SORTING

Access time is of major significance in using disk files. How can we minimize seek time, rotational latency, and the repetitive manipulation of records to be sorted?

First, we know that the tracks of a disk are organized as cylinders to avoid switching from track to track. If we control the placement of records on the disk, we can specify that each disk file be placed on a contiguous set of tracks contained within a single cylinder. In this way the cylinder may be treated exactly like a tape, and the methods useful for tapes are also useful for disk sorting.

If the initial strings are short enough to be fully contained within each cylinder, the sort-by-merge algorithms result in a minimal number of seeks. In addition, if extra output space is provided on each cylinder along with the input string's space, we realize an additional saving while performing the merge.

In most multiprogramming systems the user has little control over the order in which disk accesses are made or the placement of large files on disk. Hence we must design a sort strategy that is intrinsically optimal. To do this we note that the "cost" of merge sorting a disk file is a combination of the number of seeks and the number of strings merged. In the tree terminology we used before, the cost of sorting on disk is expressed as the number of node degrees D and the external path length E in all paths of the merge tree. We wish to minimize

$$\alpha D + \beta E$$

where α = constant of proportionality related to overhead in a seek
  β = constant of proportionality related to overhead in a merge
  D = sum over all terminal nodes of the internal node degrees along the path from terminal to root node
  E = sum over all terminal nodes of path lengths from terminal to root node

We cannot construct the minimal merge tree as we did before because of the added restraint placed on node degree. We can minimize the transmission time (time to access and copy from disk into a main memory buffer) by increasing c in the c-nary merge tree. However, we must remember that main memory is limited compared with the size of the file to be sorted.

Suppose that we wish to sort ten strings all of the same length. We show the construction of three minimal merge trees in Figure 7.14. In each of these trees we have set $\alpha = \beta = 1$ to emphasize seek time over transmission time. The tree of Figure 7.14(b) is best in this case, because its E + D value is the lowest of the three trees. Therefore the best strategy is to perform two three-way merges, one four-way merge, and then one three-way merge.

The trees of Figure 7.14 are constructed by modifying the merge trees of the previous section for c = 2, 3, 4. At first sight it would appear that a

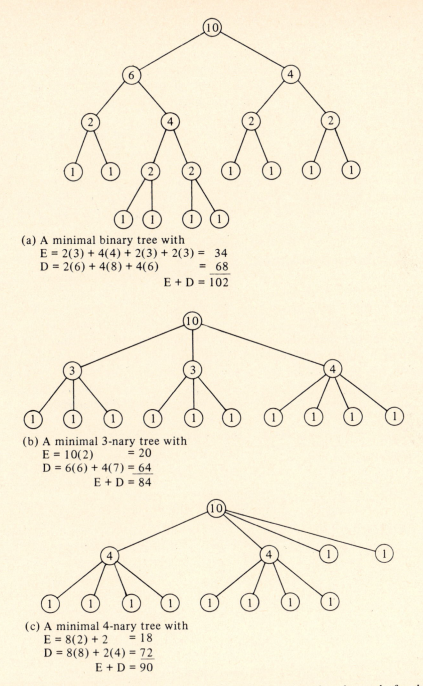

(a) A minimal binary tree with
   E = 2(3) + 4(4) + 2(3) + 2(3) = 34
   D = 2(6) + 4(8) + 4(6)   = 68
                        E + D = 102

(b) A minimal 3-nary tree with
   E = 10(2)   = 20
   D = 6(6) + 4(7) = 64
          E + D = 84

(c) A minimal 4-nary tree with
   E = 8(2) + 2   = 18
   D = 8(8) + 2(4) = 72
          E + D = 90

*Figure 7.14 Modified minimal merge trees for ten terminal nodes and of order
(a) two, (b) three, and (c) four*

Huffman tree of optimal degree satisfies the minimum E + D criteria, but this is not so.

We construct the optimal disk sort tree by calculating E + D for every tree with ten (in general n) terminal nodes, and then selecting the tree with the smallest value. Note, however, that all subtrees of an optimal tree are also optimal. We can thus recursively construct optimal disk sort trees from a relatively brief table of optimal subtrees.

Table 7.1 contains a list of optimal subtrees for disk sorts on initial strings all of the same length.

If we consult Table 7.1 for $n = 10$, we find that the optimal tree is one with three branches leaving the root node. The subtrees formed by son 1, son 2, and son 3 have degree 3, 3, and 4, respectively. To obtain the shape of the three-son subtrees, we look up the cases $n = 3$ and $n = 4$. The result is the tree shown in Figure 7.14(b).

In specific cases it may be necessary to apply a different measure of optimality. In such cases (which are the rule more often than the exception) we would construct trees appropriate to the measure of optimality.

In general, total transmission time is minimized if we let c be as large as possible in the c-nary merge tree. Increasing c also increases the amount of main memory required, so we have competing effects to consider.

A complete c-nary merge tree is optimal if the initial strings are all the same length, but if replacement-selection is used to create the initial strings, this will not generally be the case. Merge patterns for unequal-sized initial strings can be obtained from the Huffman tree (no seek time considered) or from the minimum E + D trees that are experimentally derived (no method

Table 7.1 Optimal disk sort subtrees

n	*Degree of root node of subtree*	*Filial set size* Son 1	*Son 2*	*Son 3*	*Son 4*	*Son 5*
2	2	1	1			
3	3	1	1	1		
4	4	1	1	1	1	
5	5	1	1	1	1	1
6	2	3	3			
7	3	1	3	3		
8	3	2	3	3		
9	3	3	3	3		
10	3	3	3	4		
11	3	3	4	4		
12	3	4	4	4		
13	4	3	3	3	4	
14	4	3	3	4	4	

Source: D. E. Knuth, *The Art of Computer Programming*, vol. 3, *Sorting and Searching* (Addison-Wesley Publishing Co., 1973), p. 368.

other than enumeration exists for finding minimum E + D trees when the initial strings are of different length).

Finally, methods other than merge-sorting may be useful in disk-file sorting for particular instances. The radix sort, for example, is sometimes used when the range of key values is so small that all records with a given key value exceed main memory by roughly a factor of 2.

We will state again the rule that there is no single sort algorithm that is best in all cases; the same is true for disk sorting. Know the various methods and select the one that best fits your needs.

EXERCISES

1. Perform a replacement-selection sort on the records given by the keys: 10, 20, 15, 25, 30, 12, 2. What is the number of comparisons needed to sort these keys? Assume that the sort area can hold three records.

2. Use the list of Exercise 1 to construct a binary Huffman minimal-merge tree. Compute the tree's value and then the number of comparisons needed to sort the keys.

3. Suppose that your computer installation has $t = 5$ tape units. A certain payroll application requires merge-sorting 25 strings initially of length $L_0 = 1024$. Draw a minimal-merge tree for this application. What are c and r? (*Caution:* Remember that one tape must always be free to store the merged output.)

4. Learn how to use your system's sort utility.

5. Write an algorithm for replacement selection.

6. Make Figure 7.6 into two figures showing the allocation of storage and files for the sort phase and the merge phase.

7. What considerations are involved when we wish to sort and order the data with respect to more than one key? For example, if there are two keys, then records that have the same first key must have their second key field in the correct order.

8. Simulate a balanced, cascade, and polyphase merge sort on two files: one initially of eight strings and one initially of nine strings. Assume $t = 4$ tapes. How do the methods compare?

9. Extend the table of Figure 7.11 for (a) $t = 4$ and $k = 8$, (b) $t = 5$ and $k = 4$.

10. Read the article by Shell, and perform the horizontal distribution algorithm for polyphase merge sort, when there are $t = 4$ tapes and 13 initial strings.

8
SEARCHING

8.1
Search
Effort

Searching is the process of scanning a data structure or file to find a particular atom. The efficiency of a search algorithm depends on the data structure and, in turn, the data structure depends on how the algorithm manipulates the data. Therefore it is not clear which comes first—selection of the search algorithm or design of the data structure.

Chapter 8 will concentrate on the search algorithm alone. Most of the examples use a dense-list data structure. However, the measures of performance are generally independent of the data structure. We will consider space requirements, algorithm complexity, and number of move and compare operations.

One measure of search algorithm performance is the *average search length:* the expected number of comparisons needed to locate an item in the data structure. This is a statistical measure, so some searches will take longer than others because of the data involved. The average search length is only one way to compare different algorithms.

Suppose that a dense list of Social Security numbers is stored and accessed with the frequency shown in Figure 8.1. Perhaps these are part of an employee table maintained by a company or records kept by the Internal Revenue Service. The access frequency shows the number of times each record is accessed. With these numbers it is possible to compute an average search length for various algorithms. We show only a small list in the text. The principles involved are the same whether the list is small or large.

From the table in Figure 8.1 we can estimate the probability of accessing an element (p_i) by dividing its access frequency (f_i) by the total number of accesses. Thus, for any element,

Index	Social Security numbers	Access frequency (f_i)
1	542-92-2241	100
2	381-82-1105	55
3	662-21-5503	10
4	212-55-3891	250
5	442-52-4122	125
6	588-29-9953	300

Figure 8.1 Sample list with access frequency counts

$$p_i = \frac{f_i}{\sum\limits_{i=1}^{n} f_i} \qquad \text{for } i = 1, 2, \ldots, n$$

We assume that there are n items in the table.

For the table in Figure 8.1, we can compute the following probabilities. The first step is to calculate

$$\sum_{i=1}^{n} f_i = 100 + 55 + 10 + 250 + 125 + 300 = 840$$

and then

$$p_1 = \frac{f_1}{840} = \frac{100}{840} = .119 \qquad p_4 = \frac{f_4}{840} = \frac{250}{840} = .298$$

$$p_2 = \frac{f_2}{840} = \frac{55}{840} = .065 \qquad p_5 = \frac{f_5}{840} = \frac{125}{840} = .148$$

$$p_3 = \frac{f_3}{840} = \frac{10}{840} = .012 \qquad p_6 = \frac{f_6}{840} = \frac{300}{840} = .357$$

We can define the average search length L as follows:

$$L = \sum_{i=1}^{n} c_i p_i \tag{8.1}$$

where n is number of table entries, c_i is number of comparisons needed to reach the ith entry, and p_i is probability of accessing the ith entry. Computing c_i depends on the search algorithm. In the following sections we compute L for different search methods.

EXERCISES

1. Give three measures of search effort that might be used to compare search algorithms.

2. Write a computer program to compute the average search length for any search method. Let c_i and p_i for $i = 1, 2, \ldots, n$ be inputs to your program. Given n, compute L.

3. Let $c_i = i$ and compute L for the six Social Security numbers shown in the text.

8.2
Comparison of
Search Algorithms

SEQUENTIAL SEARCH

The simplest search algorithm to understand and to program is the *sequential*, or *linear*, *search*. Whether working with files, tables, or lists, we use a similar procedure. In a table search, each entry is compared with a particular key to be located. The comparison begins with the first table element and proceeds sequentially through the table. When the matching key is found, the search returns the address or index of the item. If no match occurs, an appropriate signal is returned. A sequential search of a file begins with the first record and reads each record until it locates the record containing the desired key.

The search effort for this algorithm is computed by substituting into Formula (8.1) in Section 8.1. The number of comparisons needed to reach the first entry is one, the second entry two, and so on. In general, to reach the ith entry i comparisons are necessary. For the sequential search the average search length is

$$L = \sum_{i=1}^{n} c_i p_i = \sum_{i=1}^{n} i p_i$$

$$= 1 \cdot p_1 + 2 \cdot p_2 + 3 \cdot p_3 + \cdots + n \cdot p_n$$

For the example in Figure 8.1,

$$L = 1(.119) + 2(.065) + 3(.012) + 4(.298) + 5(.148) + 6(.357)$$

$$= 4.36$$

In other words, 4.36 comparisons are required on the *average* to locate each desired record. For this example $n = 6$ and the search method requires searching over half the table *most of the time*.

Consider the frequency counts in the table in Figure 8.1. The most frequently accessed entry is the last in the table. Suppose that the table is ordered by frequency: The most frequently accessed items are placed at the beginning. Then the value for L shows an improvement:

$$L_{\text{ordered}} = 1(.357) + 2(.298) + 3(.148) + 4(.119) + 5(.065) + 6(.012)$$

$$= 2.27$$

This nearly doubles the speed of the sequential search algorithm, but we no longer have freedom to choose the way the items are placed in the table. In many cases we may not know the frequency of access of the data. Such ordering by frequency restricts the update characteristics of the list in exchange for improved search speed. What is the limit of improvement? Can we exchange other "freedoms" of list construction for speed? Assume that we know nothing about the frequency (and hence probability) of access. The worst possible case is that all items are accessed equally often. This means that

$$p_i = \frac{1}{n} \qquad \text{for } i = 1, 2, \ldots, n$$

The average search length for the sequential search algorithm now becomes

$$L_{\text{random}} = \sum_{i=1}^{n} c_i p_i = \sum_{i=1}^{n} i\left(\frac{1}{n}\right) = \frac{1}{n} \sum_{i=1}^{n} i$$
$$= \left(\frac{1}{n}\right) \frac{n(n+1)}{2} = \frac{n+1}{2}$$

Hence total ignorance (and complete freedom of placement of items) leads to an anticipated search length equal to approximately half the length of the table.

Example What is the expected time to find an item in a table of length 1000, assuming uniform access probability and 10^{-6} second per comparison?

$$\text{Time} = 10^{-6} L_{\text{random}} = 10^{-6} \frac{1000 + 1}{2}$$

$$\approx 10^{-6} \frac{10^3}{2} \approx .0005 \text{ second}$$

BINARY SEARCH

We can improve access time if we order the data. The order can be alphabetical or numerical. When we are given a key and a request to find its record in a table (or file) of n records, we begin by looking at the middle record, that is, the item nearest location $n/2$. If the key of item $n/2$ is not the one we are looking for, we determine whether the key of item $n/2$ is larger or smaller than the given key. Because the list is ordered, we can eliminate either the first half or the second half of the total number of records from further consideration. We can repeat the comparison on our sublist and successively halve the lists to close in on the desired item.

Suppose that we are given the following list and are asked to search for the key = 47:

$$3 \quad 18 \quad 47 \quad \underline{54} \quad 65 \quad 83 \quad 94 \quad 97$$
$$\uparrow \qquad\qquad\qquad\qquad\qquad\qquad \uparrow$$

The middle record is found by noting that 54 is nearest location $n/2$. Since 54 is greater than 47, we eliminate the right half and consider

$$3 \quad \underline{18} \quad 47$$

Since 18 is at the center of the list and 18 is less than 47, we discard the left portion and are left with 47. We have thus made a match in only three comparisons!

If we begin with a list of n elements and discard half the elements with each comparison, how long will it take us to discard all the elements? If we find the element we are searching for, we stop the search and do not go through the entire list. But if the item is not there, how many comparisons must we make before determining that fact? This is equivalent to asking how many times we can successively divide n (and its quotients) by 2, and the answer is

$$c = \log_2 n$$

Another way of expressing this is $2^c = n$, where c is the number of comparisons required.

This searching technique is known as a *binary search*. It succeeds by successively halving the lists until a sublist of length one occurs. Either that sublist contains the desired item or the desired item is not in the list.

The maximum search length we can have is $\log_2 n$. If the key corresponds to the middle element, the search length is one. Suppose that the list is very long. How many entries can be located by a binary search algorithm in exactly one comparison? in exactly two comparisons? This information is summarized in the following table.

Comparison	Number of items that can be found
1	$1 = 2^0$
2	$2 = 2^1$
3	$4 = 2^2$
.	.
.	.
.	.
k	2^{k-1}

But what is the *average* search length? Let us assume that $n = 2^k - 1$, and that the probability of searching for each key is equally likely. Then statistically speaking, we would expect to have searched for each item exactly once after processing n requests. The average search length is the total number of comparisons divided by n:

$$L = \frac{\text{total comparisons}}{n}$$

To calculate the total number of comparisons required to access all items in the list, note that the middle element contributes one comparison to the total. Each of two elements contributes two comparisons; each of four elements contributes three comparisons; each of eight elements contributes four comparisons; and so on. Thus

$$L_{\text{binary}} = \frac{\sum_{i=1}^{k} i\, 2^{i-1}}{n}$$

Without going through the derivation, we evaluate L as

$$L_{\text{binary}} = \frac{n+1}{n} \log_2 (n+1) - 1$$

Since not every list has length $2^k - 1$, and our other assumptions introduce errors into the above analysis, it is common to accept the following approximation, provided that n is large ($n > 50$):

$$L_{\text{binary}} \approx \log_2 (n+1) - 1$$

The exact formula for a list of arbitrary length n is given by the formula[1]

$$L_{\text{binary}} = \log_2 (n+1) - \frac{1}{n} \left(2^{[\log_2(n+1)]} - [\log_2 (n+1)] - 1 \right)$$

where the brackets [ ] mean to take the integer part of the contents of the brackets.

In summary, the binary search algorithm owes its search performance to the loss in flexibility incurred by ordering the list. The price for rapid access is the difficulty in keeping the list ordered after repeated updates. Thus, because the list changes frequently with insertions and deletions, the cost to keep it ordered may outweigh its fast search time.

BLOCK SEARCH

For rapidly changing files that are not accessed too often, there is a compromise between the sequential search and the binary search. This search method

[1] For a fuller treatment of this formula, see C. V. Ramamoorthy and Y. H. Chin in References.

is the block search. The data is divided into blocks; within each block we do not have to order the data, but we must keep the blocks in order. This means that all items in the first block are less than all items in the second block, and so on, assuming that the list is in ascending order. If the list were in descending order, all items in the first block would be greater than all items in the second block, and so on.

To facilitate the search we must know the largest value in each block (if the list is in ascending order). Searching the list means comparing the key with the largest value in each block until the key is less than some largest value. Having located the block that should contain the key, we use a sequential search within the block. If the key is found, we return its location. If the key is not found, we return an appropriate value to signal that the key is not in the list.

For example, suppose that we have the list pictured in Figure 8.2. The column labeled MAXVAL contains the indexes of the largest value in each block. Each block contains five values. To search for the value 60, we first compare it with 22 (largest value in block 1), then with 44 (largest value in block 2), and finally to 74 (largest value in block 3). Thus, we determine that 60 is in the third block, if it is in the list at all. Next we search the third block sequentially and find 60 in location 12.

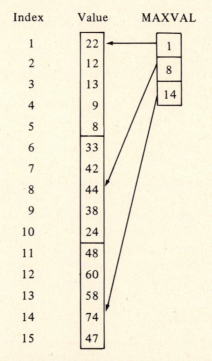

Figure 8.2 List prepared for block search

The average search length for a block search reveals that it is an improvement over sequential search, but is not nearly as efficient as the binary search. The block search involves two searches: one to find the correct block and one to find an item within the block. Thus, the average search length

$$L_{block} = L_b + L_w$$

where L_b is the average search length for blocks and L_w is the average search length within blocks. Let us say that the list contains n elements, s elements per block, and b blocks, where $b = n/s$. If the probability of searching for an element is the same as that of searching for any other element, then the probability of accessing any block is $1/b$ (or s/n) and

$$L_b = \sum_{i=1}^{b} (i)\left(\frac{1}{b}\right)$$

The probability of accessing any item within the block is $1/s$. Only $s - 1$ items need be compared, because one in each block is used to search from block to block. Thus,

$$L_w = \sum_{i=1}^{s-1} (i)\left(\frac{1}{s}\right)$$

Then the average search length for any item is

$$
\begin{aligned}
L_{block} &= \sum_{i=1}^{b} (i)\frac{1}{b} + \sum_{i=1}^{s-1} (i)\frac{1}{s} \\
&= \frac{1}{b} \sum_{i=1}^{b} i + \frac{1}{s} \sum_{i=1}^{s-1} i \\
&= \frac{1}{b} \frac{b(b+1)}{2} + \frac{1}{s} \frac{(s-1)s}{2} \\
&= \frac{b+1}{2} + \frac{s-1}{2} = \frac{b+s}{2} \\
&= \frac{1}{2}\left(\frac{n}{s} + s\right) = \frac{n+s^2}{2s}
\end{aligned}
$$

Thus, the average search length depends not only on n but also on s, the number of elements in each block. For a given list, n is constant but s may vary. To minimize L_{block} we use some results from calculus. First we set the derivative of L_{block} with respect to s equal to zero and solve for s:

$$\frac{dL_{block}}{ds} = \frac{1}{2}\left(1 - \frac{n}{s^2}\right) = 0$$

Then $s = \sqrt{n}$ provides a minimum average search length. Substituting $s = \sqrt{n}$ in the formula for L_{block}, we find that the average search length L^*_{block} is

$$L^*_{block} = \frac{n + (\sqrt{n})^2}{2\sqrt{n}} = \frac{2n}{2\sqrt{n}} = \sqrt{n}$$

For example, a block search on a list of 10,000 entries requires on the average 100 comparisons to retrieve an item. This assumes that there are 100 entries/block and a total of 100 blocks. A binary search of 10,000 entries would take an average of 13.3 comparisons; a sequential search would take an average of 5000.

BINARY SEARCH TREE

Another alternative that allows both rapid search and ease of update is the binary search tree. In Chapter 5 it was referred to as the binary sort tree because we wanted to convert a list into a tree structure in such a way that the nodes would be ordered, if scanned in LNR order. Updating and searching the tree are simplified because of the way it is defined. Recall that in a binary tree each node is a terminal node or it has at most two immediate successors; the successors are labeled left and right successors. An LNR binary search tree is constructed so that a node's immediate left successor is less than the node, and its immediate right successor is greater than the node. Thus, all the elements on a node's left subtree are always less than all the elements on the node's right subtree. Searching the tree for a particular item is a binary-type search. We begin at the root and eliminate either the left or right subtree, depending on the result of the comparison. We continue the search until either the item is found or a terminal node is reached.

For example, Figure 8.3 illustrates the LNR binary search tree constructed from the list of numbers {4, 2, 3, 6, −1, 12, 8}. Recall that a different ordering of the same numbers would produce a different tree structure because nodes are added in the order that they are found in the list.

In Chapter 5 we discussed the addition of nodes to the tree. We now turn

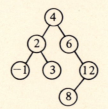

Figure 8.3 A binary search tree

our attention to the problem of deleting nodes from this structure that gives us both update flexibility and search speed. After a deletion, the tree must remain a binary search tree so that other insertions or searches may occur with ease. Figure 8.4 shows before and after pictures of various binary search trees. See if you can determine how the shape of the tree changes.

Visually, the basic rule of deletions involves replacing the deleted node with a terminal node. The terminal node chosen is the rightmost terminal node in the deleted node's left subtree. In this way the tree remains a binary search tree. If the deleted node is already terminal, no replacement occurs. If the deleted node's left subtree has no right branch, then we move the deleted node's left subtree up. If the deleted node has no left subtree but does have a right subtree, then we replace the deleted node with its right successor and move the right subtree up.

(a) Before deletion of 2   (a) After deletion of 2

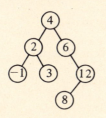   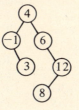

(b) Before deletion of 5   (b) After deletion of 5

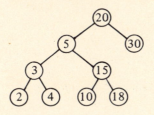   

(c) Before deletion of 6   (c) After deletion of 6

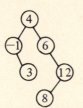

Figure 8.4 Deletion of nodes from binary search trees

Since this may seem confusing, refer to the examples in Figure 8.4. In Figure 8.4(a), node 2 is to be deleted. Its left subtree has no right subtree, so the left subtree moves up. The pointers of the −1 node are adjusted so that its right successor is 3 and its left successor is nil. In Figure 8.4(b) node 5's left successor is 3 and 3's rightmost terminal node is 4. Thus 4 moves up to 5's position and all other nodes stay in their same position. Figure 8.4(c) illustrates the case when the deleted node has a right subtree but no left subtree. Node 6's right subtree headed by 12 moves up to 6's position.

Each time a deletion takes place, we must search the left subtree emanating from the deletion node to find the rightmost node, or take the appropriate action if there is no left subtree with a rightmost node. The general LNR_DELETE algorithm follows.

Algorithm LNR_DELETE

1. Comments: Delete a node D from an LNR tree.
    Input: D   Node to delete
    Output: none   Unless an error occurs
    Variables: R   Pointer to replacement node that occupies tree position
                                formerly occupied by D
                 LEFT   Left successor of node
                 RIGHT   Right successor of node
                 PARENT   Parent of node

2. Search tree and locate node D and node D's PARENT.
        If D is not in the tree, then
            Output 'Error'
            EXIT LNR_DELETE

3. If LEFT of D is nil, then
        If RIGHT of D is nil, then
            R ← nil
        Else
            Move up right subtree:
                R ← RIGHT of D
    Else
        If RIGHT of D is nil, then
            Move up left subtree:
                R ← LEFT of D
        Else
            Search down right subtree of LEFT of D until reaching a terminal node;
            Set R ← terminal node
            Set PARENT of R ← nil

4. Update parent pointer:
        If D is LEFT of PARENT, then
            LEFT of PARENT points to R

Else
    RIGHT of PARENT points to R

5. Update R's pointers:
    LEFT of R ← LEFT of D
    RIGHT of R ← RIGHT of D

6. End LNR_DELETE

One possible application of a binary search tree is the storage of an airline's rapidly changing reservations file. Assume that each flight has its own binary search tree containing the names and telephone numbers of passengers. Making a reservation creates an additional node on the tree. Canceling a reservation deletes a node from the tree. Last names determine the ordering. A name appears on a node's left subtree if it precedes the node alphabetically; a name appears on the right subtree if it follows the node alphabetically.

Whether adding or removing a name or just checking a particular person's confirmation for a flight, we must search the tree. Can we estimate the average search length for a typical binary search tree? Here we assume that reservations and cancellations occur randomly. It can be shown (although we will not do it here) that for a binary search tree, the average search length is bounded by

$$L_{tree} \leq 1.4 \log_2 (i - d)$$

where we begin with an empty tree and make i random insertions and d random deletions.

For the airline's binary search tree, if we assume 250 passengers and 10% cancellations, then the average search length is

$$L_{tree} \leq 1.4 \log_2(250 - 25)$$

$$= 1.4 \log_2(225)$$

$$= 10.94$$

This means that the search length for the airline's binary search tree is approximately 11. We can only estimate L_{tree} because the tree is constantly changing in size. The more insertions there are, the larger the average search length.

The binary search tree organization does compare favorably in average search length with the binary search. However, the memory requirements are nearly twice that of a dense list organization because we must provide room for the pointers.

Another problem of the binary search tree is that the tree grows in the

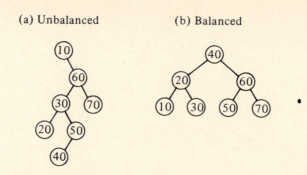

Figure 8.5 Balanced versus unbalanced binary search tree

order in which insertions occur. Compare the two binary search trees in Figure 8.5. The tree in Figure 8.5(a) results from the list {10, 60, 30, 70, 20, 50, 40}; the tree in Figure 8.5(b) results from {40, 20, 60, 30, 70, 50, 10}. The binary tree (a) is not balanced because it is not symmetric, while the binary tree (b) is symmetric and therefore balanced. An unbalanced binary search tree results in longer search lengths and slower updates. Is there some way to balance a binary search tree?

Algorithms for balancing binary search trees are given by Martin and Ness (see References) and will not be discussed here. Also note that binary search trees are a special kind of AVL (Adel'son, Vel'sky, and Landis) tree. These trees have been studied extensively for their searching and sorting properties.

SUMMARY

To summarize the searching algorithms, consider the following: Programming a dense list sequential search is simple and is perhaps the "best" method to use on short lists. The average search length is proportional to the length of the list.

The binary search is rapid because it requires very few comparisons. However, it will work only on an ordered list. Therefore it is best suited to a dense list structure that does not change or that changes slowly. Updates to an ordered list are slow because they must preserve order.

A block search can be optimized to provide moderately fast searches and moderately flexible update capability. The storage requirements and programming simplicity of a block search make it an alternative to a binary search tree.

A binary search tree allows rapid lookup and ease of insertion and deletion. It requires a tree structure, which needs extra memory for pointers. Thus memory requirements increase in exchange for speed and flexibility.

The following table summarizes the features and advantages of the search algorithm presented in this section.

Features / Algorithm	Search length	Memory space	Programming complexity	Main advantages
Sequential	$\dfrac{n + 1}{2}$	n	Simple	Programming
Block	$\sqrt{n}$	$\sqrt{n} + b$	Moderate	Programming and updating
Binary	$\log_2 (n + 1) - 1$	$n + 2$	Moderate	Search length (speed)
Search tree	$1.4 \log_2 (i - d)$	$\approx 2n$	Difficult	Search length (speed) and updating

EXERCISES

1. Write an algorithm for one of the search algorithms.

2. Write a program for one of the search algorithms and try it on different data sets. Compute the theoretical average search length and compare it with the empirical average produced by the program runs.

3. What assumptions were made about the access frequency for each average search length calculation?

4. What is the expected number of moves needed to insert a new item in the list searched with a binary search? What are your assumptions?

5. Suppose that a table is ordered by access frequency and that a copy is ordered lexicographically. Which table can be searched faster? Qualify your answer by saying something about the access frequencies and the maximum search lengths.

6. How many different binary search trees can be constructed from the numbers 10, 20, 30, 40, 50? Compute the expected search length for each tree and show which one is smallest.

7. What are the optimal block sizes for a block search on a list of 144 items? Suppose that the same list has blocks of eight items each. What is the average search length now? What is the percentage loss in average search efficiency?

8. Suppose that the block search uses a binary search to determine the correct block and then a sequential search within the block. What is the average search length?

9. Write programs for each search algorithm and compare their performance, space requirements, and programming complexity.

10. Design an NLR binary search tree deletion algorithm. How does this tree differ from the LNR search tree? How is its data ordered?

11. Implement the LNR deletion algorithm.

8.3
Hashing

We have seen how the average search length can be affected by keeping the data in order. It is conceivable that other methods might yield even better search characteristics. The key-to-address transformation, or hashing[2] (see also Chapter 6 and Section 10.11), is one in which the data need not be ordered but which provides high-speed searching capability.

Suppose that a file is to be designed for which rapid insertion, deletion, and lookup are of paramount importance. If we are willing to sacrifice storage, hashing has a remarkable potential. The basic idea behind hashing is that some operation is performed upon the key to be located. That operation determines a value, which is used as an index or address into a table. It is hoped that this locates the key's data in just one trial. The number of trials required to locate an entry depends on such factors as the hashing algorithm, the data being searched, the size of the table, as well as other considerations. In this section we will go into some of the more common hashing operations before analyzing its search characteristics.

DIGIT ANALYSIS

Assume that we have a list of Social Security numbers as shown in Figure 8.6. It would be convenient to use the numbers themselves as addresses. Therefore 542-42-2241 is stored in memory location 542422241, 542-81-3678 in 542813678, and so on. This means that the few numbers in our example would need close to a billion locations. Unquestionably this approach is fast: The maximum search length is one. But let us see how we can eliminate some of the wasted space from the example.

Suppose that we have only 1000 memory locations available and want to define an address for each of the numbers in Figure 8.6. The first three digits are all the same, so we eliminate them immediately. The next six digits would be all right, but we really need only three (000-999). How can we intelligently fit these six Social Security numbers into 1000 locations? A closer look at the

542-42-2241
542-81-3678
542-22-8171
542-38-9671
542-54-1577
542-88-5376
542-19-3552

Figure 8.6 Social Security numbers for hashing

[2] We will use the terms hashing and key-to-address transformation interchangeably.

numbers reveals a preponderance of number 7 in the second column from the right. Since the digits in that column do not help spread the numbers out, we will eliminate that column also. We continue to examine the key set for other columns with nonrandom patterns. If none are obvious, we arbitrarily eliminate some others, for example, 4, 5, and 8. The key-to-address transformation (pick columns 6, 7, and 9) for the example (Figure 8.6) appears in Figure 8.7.

We are illustrating the principle that only digits that have the most uniform distribution are selected. Digits with the most nonuniform (skewed) distribution are deleted from the key until the desired number of digits is obtained. A program can be written to perform digit analysis, but this requires advanced knowledge of the keys. A change in the key set requires a new digit analysis, so this method does not meet our requirement of rapid insertion and deletion capability.

FOLDING

It often happens that the key is longer than the number of digits in the address range. If the distribution of digits in the key is roughly uniform, the choice of which digits to ignore is rather arbitrary. One alternative (on which there are countless variations) is to select some columns arbitrarily and add or exclusive-or the remaining ones to the chosen ones. Figure 8.8 demonstrates two methods of folding Social Security numbers into an address space of four digits (0000-9999). The intention is to use all the digits to help scramble the addresses uniformly across the address space. If the sum produces overflow digits, they are usually discarded. Folding is frequently combined with other hashing methods.

DIVISION

One of the earliest and easiest key-to-address transformations is division by the number of possible addresses. For example, division of any number by n always leaves a remainder in the range of zero to $n - 1$. The transformation

Key	Address
542-42-2241	221
542-81-3678	368
542-22-8171	811
542-38-9671	961
542-54-1577	157
542-88-5376	536
542-19-3552	352

Figure 8.7 Key-to-address transformation on Social Security numbers

(a) Before folding, creases are selected

Social Security number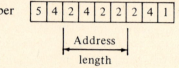

(b) Fold-boundary with sum

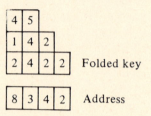

(c) Fold-shifting with sum

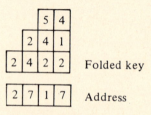

Figure 8.8 Fold-boundary and fold-shifting transformations

of a number to its remainder after division by a fixed number is the central idea in modular arithmetic. For example, we say that

$$18 = 0 \bmod 6 \qquad 19 = 1 \bmod 6 \qquad 20 = 2 \bmod 6$$

because division by 6 leaves remainders of 0, 1, and 2, respectively. Most hashing experts suggest division by a prime number, slightly smaller than the table size. We will see that there are advantages in selecting a prime number for table length. Figures 8.9 (pp. 196–197) and 8.10 (p. 198) give lists of primes.

PSEUDORANDOM

The division algorithm is reminiscent of the way multiplicative pseudorandom number generators work. Such a generator typically has the form

$$y = (ax + c) \bmod p$$

Replacing x by the key produces y for its address. Typically p is a power of 2 of the form 2^n and $c = -1$. If we select $a = 1 \bmod 4$, it is possible to generate every number from zero to $p - 1$ exactly once by starting with any number in the same range. To use this formula as a hashing function, we let

$$\text{Address} = (a \cdot \text{key} + c) \bmod p$$

$$a = 1 \bmod 4 \qquad c = \pm 1 \qquad p = 2^n = \text{table size}$$

For example, if $a = 5$, $c = 1$, $p = 2^{10} = 1024$, and our key is 3386, then

$$\text{Address} = (5 \cdot 3386 + 1) \bmod 1024$$
$$= (16931) \bmod 1024 = 547$$

When we must use hashing on character data, we use the numeric value of the characters as if they were a number. For example, the word 'CAT' in EBCDIC coding is C3C1E3 (hexadecimal notation), which is 12829155 in decimal.

Pseudorandom hashing is often used as a standard against which the randomizing effect of other hash algorithms is measured. Other methods of hashing that may offer advantages in certain cases have been proposed. You may invent a personalized hashing function of your own. However, the division technique is fairly simple and is quite effective for most applications.

RESOLVING COLLISIONS

Since the purpose of hashing is to squeeze a potentially large key space into a small address space, it is inevitable that *collisions* will occur. That is, two or more keys are transformed into the same address. Before a complete data structure based on hashing can be implemented, we must decide what to do when collisions occur. Once we make that decision, insert, delete, and search algorithms are all basically the same.

Suppose that we are in the process of building a table using some hashing function. We are given a record and use the hash function on the key to generate an address. What can we do if something is already stored at that address? Or suppose that we are searching for a particular record; we hash its key to give an address, but the record stored at that location is not the one we wanted. Where do we look next?

There are many methods for solving these problems. For example, if the computed address is full, check the next address, then the next, and so on, until a vacant spot is found for the new record. Searching is done the same way. If the computed address does not contain the desired record, check the next record, then the next, and so on. This is a simple scheme but it has some drawbacks. If several keys hash to the same address, the sequential searching

2	181	431	683	977	1277	1567
3	191	433	691	983	1279	1571
5	193	439	701	991	1283	1579
7	197	443	709	997	1289	1583
11	199	449	719	1009	1291	1597
13	211	457	727	1013	1297	1601
17	223	461	733	1019	1301	1607
19	227	463	739	1021	1303	1609
23	229	467	743	1031	1307	1613
29	233	479	751	1033	1319	1619
31	239	487	757	1039	1321	1621
37	241	491	761	1049	1327	1627
41	251	499	769	1051	1361	1637
43	257	503	773	1061	1367	1657
47	263	509	787	1063	1373	1663
53	269	521	797	1069	1381	1667
59	271	523	809	1087	1399	1669
61	277	541	811	1091	1409	1693
67	281	547	821	1093	1423	1697
71	283	557	823	1097	1427	1699
73	293	563	827	1103	1429	1709
79	307	569	829	1109	1433	1721
83	311	571	839	1117	1439	1723
89	313	577	853	1123	1447	1733
97	317	587	857	1129	1451	1741
101	331	593	859	1151	1453	1747
103	337	599	863	1153	1459	1753
107	347	601	877	1163	1471	1759
109	349	607	881	1171	1481	1777
113	353	613	883	1181	1483	1783
127	359	617	887	1187	1487	1787
131	367	619	907	1193	1489	1789
137	373	631	911	1201	1493	1801
139	379	641	919	1213	1499	1811
149	383	643	929	1217	1511	1823
151	389	647	937	1223	1523	1831
157	397	653	941	1229	1531	1847
163	401	659	947	1231	1543	1861
167	409	661	953	1237	1549	1867
173	419	673	967	1249	1553	1871
179	421	677	971	1259	1559	1873

Figure 8.9 List of primes

that follows can become too costly. This effect is called *primary cluster-ing,* and it occurs when the hashing function does not spread the addresses uniformly throughout the address range.

Another alternative is an extension of the pseudorandom hashing tech-nique. If the computed address collides with a table occupant, a subsequent location is interrogated by computing a number called an *offset,* which is added to the original location. The various methods of handling collisions differ only in the manner of calculating offsets.

In pseudorandom probing, the offset x_i is computed from the following formula in which n is the table length:

$$x_i = (ax_{i-1} + c) \bmod n$$

1877	2203	2503	2801	3167	3469	3779
1879	2207	2521	2803	3169	3491	3793
1889	2213	2531	2819	3181	3499	3797
1901	2221	2539	2833	3187	3511	3803
1907	2237	2543	2837	3191	3517	3821
1913	2239	2549	2843	3203	3527	3823
1931	2243	2551	2851	3209	3529	3833
1933	2251	2557	2857	3217	3533	3847
1949	2267	2579	2861	3221	3539	3851
1951	2269	2591	2879	3229	3541	3853
1973	2273	2593	2887	3251	3547	3863
1979	2281	2609	2897	3253	3557	3877
1987	2287	2617	2903	3257	3559	3881
1993	2293	2621	2909	3259	3571	3889
1997	2297	2633	2917	3271	3581	3907
1999	2309	2647	2927	3299	3583	3911
2003	2311	2657	2939	3301	3593	3917
2011	2333	2659	2953	3307	3607	3919
2017	2339	2663	2957	3313	3613	3923
2027	2341	2671	2963	3319	3617	3929
2029	2347	2677	2969	3323	3623	3931
2039	2351	2683	2971	3329	3631	3943
2053	2357	2687	2999	3331	3637	3947
2063	2371	2689	3001	3343	3643	3967
2069	2377	2693	3011	3347	3659	3989
2081	2381	2699	3019	3359	3671	4001
2083	2383	2707	3023	3361	3673	4003
2087	2389	2711	3037	3371	3677	4007
2089	2393	2713	3041	3373	3691	4013
2099	2399	2719	3049	3389	3697	4019
2111	2411	2729	3061	3391	3701	4021
2113	2417	2731	3067	3407	3709	4027
2129	2423	2741	3079	3413	3719	4049
2131	2437	2749	3083	3433	3727	4051
2137	2441	2753	3089	3449	3733	4057
2141	2447	2767	3109	3457	3739	4073
2143	2459	2777	3119	3461	3761	4079
2153	2467	2789	3121	3463	3767	4091
2161	2473	2791	3137	3467	3769	4093
2179	2477	2797	3163			

Figure 8.9 (continued)

The first time a collision occurs, we compute an offset of x_1 from the preceding formula, assuming that x_0 = key. The next address interrogated is the hash address plus x_1. If that produces a collision, we compute another offset x_2 and add it to the hash address. If needed, several offsets might be computed before we meet with success (or failure). Simpler random-number generators are possible, but they do not guarantee that every location of the table will be scanned before the original location is scanned again. The table length n must be a power of 2, and a and c must be selected properly to ensure a full table scan.

The pseudorandom probe requires lengthy calculations and serves only to spread primary clusters over the entire table. If several keys hash to the same address, and if the method of computing the offset traces out the same search sequence for these synonyms, the result is called *secondary clustering*.

n	2^n	$(4k + 3)$ prime $< 2^n$
3	8	7
4	16	11
5	32	31
6	64	59
7	128	127
8	256	251
9	512	503
10	1024	1019
11	2048	2039
12	4096	4091
13	8192	8191
14	16384	16363
15	32768	32719
16	65536	65519
17	131072	131071
18	262144	262139
19	524288	524287
20	1048576	1048571
21	2097152	2097143
22	4194304	4194287
23	8388608	8388587
24	16777216	16777207

Figure 8.10 List of primes near a power of 2 and of form 4k + 3

Several investigators have noticed the buildup of secondary clustering and have proposed methods for eliminating it; these methods use additional information about the key. J. R. Bell proposes a variable offset that changes from one synonym to another. His *quadratic quotient search* employs a prime division hashing function and sets the offset equal to the quotient obtained after division. Thus the original probe is

$$\text{key mod } n$$

and the ith offset is given by

$$qi^2$$

where $q = \text{key}/n$ and n is a prime of the form $4k + 3$. This technique guarantees a scan of only half the table, but it eliminates primary and secondary clustering. It is possible to adapt this technique to get a full table scan by computing both R and $(n - R) \bmod n$, where R is the key mod n.

The quadratic quotient hash eliminates primary and secondary clustering because it produces a different offset for synonyms. The fact that a quadratic offset is computed is of secondary importance. A linear offset that uses

q = (quotient of prime division) also eliminates clustering and is fast. The linear quotient hash search algorithm is given below:

Algorithm LQHASH

1. Comments: Linear quotient hash algorithm. A table with N entries contains data to be located. Each data entry has a unique key.
    Input: KEY   Key to be located
    Output: R   Index of table entry containing KEY, if found
                 'Key not in table'   If not found
    Variables: N   Number of entries in table; N is a prime of form $4k + 3$
                 TABLE   Table to be searched
                 Q   Quadratic quotient
                 I   Probe counter

2. Initialize index and quadratic quotient:
    R ← KEY mod N
    Q ← KEY / N

3. If Q mod N is 0, then
    Set Q to some constant > 0

4. Set probe counter to 1
    I ← 1

5. While I < N repeat the following:
    If TABLE[R] contains the KEY, then
        EXIT LQHASH
    Else
        If TABLE[R] is empty, then
            Output 'Key not in table'
            R ← −1
            EXIT LQHASH
        Else
            Prepare to continue search:
                I ← I + 1
                R ← (R + Q) mod N

6. Output 'Key not in table'; set R ← −1

7. End LQHASH

Inserting a record into a table using a linear quotient hash algorithm is similar to the search algorithm. The table is initially built with all its entries marked empty. This may be accomplished by setting everything to zero or some other value denoting a nil entry. The insert algorithm uses the same initial steps, 2 to 4 as Algorithm LQHASH. Steps 5 and 6 are replaced by the following:

5. While I < N repeat the following:
      If TABLE[R] is empty, then
          Set TABLE[R] to the record to be inserted
          EXIT LQHASH
      Else
          Continue to look for an empty location:
              I ← I + 1
              R ← (R + Q) mod N

6. Output 'Key cannot be inserted'

When one is deleting a record from a table built with linear quotient hashing, it is not sufficient simply to locate the record and mark its location empty. Since that record may not be the only record with hash index R, we must have some way to indicate that the record is not there; but a subsequent search should continue looking for other records with that hash value. To do this, we may want to include a flag to denote that the record is followed by other records with the same hash value. This of course requires that we modify our insert algorithm to set this flag and modify our search algorithm to use it.

AVERAGE SEARCH LENGTH

You were promised that hashing is fast. To prove this we analyze the average search length needed to access a record. The secret to scatter storage is never to allow memory to become full. As memory fills up, the average search length increases.

Let us assume that there are n locations in memory, but only k of them are filled. Then the loading factor l is given by the relation

$$l = \frac{k}{n}$$

where we assume that $k < n$. In addition, the probability of a collision is l, which is also the probability that an extra probe will have to be made to locate the record.

When no probes have been made, the probability of making an additional probe is 1. The probability of a collision on the ith probe is l^i, and the probability of locating a record in exactly i probes is $l^{i-1}(1 - l)$. The average search length is

$$L_{hash} = \sum_{i=0}^{n} i l^{i-1}(1 - l)$$

$$\leq \sum_{i=0}^{\infty} i l^{i-1}(1 - l) = \frac{(1 - l)}{(1 - l)^2}$$

$$= \frac{1}{1 - l}$$

Observe that this formula expresses the search length as a function of the loading factor l and not as a function of the number of records. For a fixed number of records k, we can make the loading factor l close to zero by making n larger; $l = k/n$. The penalty of course is the $n - k$ memory locations that are not used. This assumes that each record occupies one memory location. If a record occupies more location, even more space is wasted.

Example   A file is to be designed using a hashing function. The loading factor is limited to 3/4. There are 30,000 records in the file. How many record locations will be required?

$$l = \frac{k}{n} \qquad \frac{3}{4} = \frac{30,000}{n}$$

$$n = (30,000)\left(\frac{4}{3}\right) = 40,000$$

What is the expected search length for each record?

$$L \approx \frac{1}{1-l} = \frac{1}{1-\dfrac{3}{4}} = 4$$

Suppose that there are 90,000 records. How much space is required and what is the expected search length?

   The formulas above express search effort for a static table. The greatest advantage of a scatter table lies in its updatability. Hash tables fluctuate over a range of $0 \le k < n$ as they are built. What happens to our estimate of average search length when l is allowed to vary? Assume that a scatter storage table grows from $l_{init} = 0$ to $l = l_{max}$. Then the estimated search effort needed to insert and delete items in this table is

$$L = \frac{1}{l_{max} - l_{init}} \int_{l_{init}}^{l_{max}} \frac{dl}{1-l} = -\left(\frac{1}{l_{max}}\right) \ln (1 - l_{max})$$

This assumes no clustering at all.

PERFECT HASHING

Perfect hashing was first posed as a problem by Knuth. Given a static set of keys, is there a hashing function that will give a unique hash value for each key and guarantee access to each atom with exactly one probe? Such a hashing function is called a *perfect hashing function*. Furthermore, the loading factor should be as high as possible, preferably 1.0. It is particularly desirable to have such a function whenever we have a set of items frequently used which may or may not point to other pertinent information. For example, a compiler needs to recognize the reserved words of a source program. A language such as COBOL has several hundred reserved (key) words, so a hashing technique is almost essential in order for the compiler to determine quickly whether or not a particular word is a reserved word. Certain key words such as READ or ADD may initiate special processing, since they cause the generation of executable instructions. Others, such as VALUE, imply the initialization of data areas. In this section we will present several perfect hashing algorithms and hope that you may be motivated to try to describe other algorithms.

Sprugnoli (see References) gives two types of perfect hashing functions: the *quotient-reduction method* and the *remainder-reduction method*. The quotient-reduction method uses the formula

$$h(k) = \lfloor \frac{k + s}{N}$$

where $\lfloor$ means "truncate to the nearest integer," k is the key being hashed, and s and N are integers. For a given set of keys, the problem is to find s and N such that for every key, $h(k)$ is unique. In his paper, Sprugnoli gives an algorithm for determining s and N once the set of keys is known. Lewis (see References) shows ways of improving that algorithm and an alternative heuristic approach.

The remainder-reduction method hashing formula is

$$h(k) = \lfloor \frac{(d + kq) \bmod M}{N}$$

where $\lfloor$ means "truncate to the nearest integer," k is the key being hashed, and d, q, M, and N are integers. Sprugnoli also gives algorithms for finding the integer constants in the hashing formula. The remainder-reduction algorithm works well on keys that are not uniformly distributed. Taking $(d + kq)$ mod M helps scramble the keys and makes them more evenly distributed. The quotient-reduction method has no mod M operation and hence works better on uniformly distributed keys.

As an example showing a remainder-reduction hashing algorithm, we take the 12 months of the year, each given as its three-digit abbreviation. We think of the month in its character form; in storage, the characters can be used as binary numbers. Figure 8.11 lists the months and the decimal value of the second two characters of the abbreviation coded in EBCDIC. Including the first character in the value is not necessary, since there are only 12 keys and the extra character only makes the key value larger.

Sprugnoli gives the constants for the remainder-reduction algorithm of the 12 keys. The values are $d = 2304$, $q = 256$, $M = 23$, and $N = 2$. Taking the keys and performing the hashing algorithm yields the results in Figure 8.11. Note that this function yields values 0 through 11. The hash table is as small as possible: Its loading factor is 1.0.

Month	Decimal	Hash	Month	Decimal	Hash
JAN	49621	5	JUL	58579	10
FEB	50626	6	AUG	58567	4
MAR	49625	0	SEP	50647	3
APR	55257	7	OCT	50147	1
MAY	49640	11	NOV	55013	9
JUN	58581	2	DEC	50627	8

Figure 8.11 Hash values for twelve months

Another method for perfect hashing functions is given by Cichelli (see References). His hash function is independent of the character coding scheme (EBCDIC or ASCII) for a particular machine. Its formula is:

$$h(k) = \text{length of } k + \text{associated value of } k\text{'s first character} + \text{associated value of } k\text{'s last character}$$

where k is the key being hashed. For a particular set of keys, we must compute the associated values for the characters. We will not present that algorithm, but will leave it as a reference for the interested reader.

The approach when applied to Pascal's reserved word list can produce a hash table of size 36. Figure 8.12 lists the reserved words and their corresponding hash values. The characters' associated values used in the hash function are the following: A = 11, B = 15, C = 1, D = 0, E = 0, F = 15, G = 3, H = 15, I = 13, J = 0, K = 0, L = 15, M = 15, N = 13, O

Reserved Word	Hash Value	Reserved Word	Hash Value
do	2	record	20
end	3	packed	21
else	4	not	22
case	5	then	23
downto	6	procedure	24
goto	7	with	25
to	8	repeat	26
otherwise	9	var	27
type	10	in	28
while	11	array	29
const	12	if	30
div	13	nil	31
and	14	for	32
set	15	begin	33
or	16	until	34
of	17	label	35
mod	18	function	36
file	19	program	37

Figure 8.12 Pascal's reserved words and hash values

= 0, P = 15, Q = 0, R = 14, S = 6, T = 6, U = 14, V = 10, W = 6, X = 0, Y = 13, and Z = 0.

Remember that perfect hashing is specially designed for a static set of keys known in advance, so the time-consuming problem of determining the perfect hash function need be done only once. Unlike the search time for other hashing schemes, the search time for perfect hashing is 1, regardless of the loading factor. In fact, we want the loading factor as large as possible, namely 1.0. If it is not possible to have a loading factor of 1.0, then a perfect hashing function with the maximal loading factor is called a *minimal perfect hashing function*.

Although we have separated the chapters on files and searching, it should be evident that the topics are interrelated. A sequential file needs a sequential search; an indexed file requires something similar to a block search; and so on. Although we have given figures for average search length, other aspects of file design, such as total memory cost, may take precedence in the choice of the overall design. At this stage file design is still an art rather than a science and often reflects the designer's personal preferences or the hardware's limitations rather than a rational choice of alternatives.

EXERCISES

1. Define the following terms and tell how they relate:
    (a) Synonym, (b) collision, (c) primary clustering, (d) secondary clustering, (e) key-to-address transformation.

2. Design a search technique that uses buckets (see Chapter 6). It should compare to hashing in the same way that the block search compares to a sequential search. Compute the average search length.

3. Write an algorithm to perform file maintenance for a scatter storage file. It should provide for insertion, deletion, and search capabilities.

4. Write a program to implement the algorithm in Exercise 3. Test it with different sets of data and compute the average search length. Compare this length against the theoretical value.

5. Suppose that the Highway Patrol wants to design a scatter storage file so that any record can be retrieved in an average of two probes. If they expect a total of 100,000 records, how much space must be provided for the file?

6. Write a general-purpose hash insert algorithm that will work with any hashing function.

7. Write a general-purpose hash delete algorithm that will work with any hashing function.

8. Write a general-purpose hash search algorithm that will work with any hashing function.

9. Update the linear-quotient search algorithm so that it uses the delete flag.

10. Choose a small set of key values and design your own perfect hashing function.

11. Choose a set of keys and a hashing function. Vary the size of the hash table and compute the loading factor for the various table sizes. Plot your results.

12. Choose a set of keys and a size for a hash table. For various hashing functions, compute each function's loading factor. Choose other table sizes and compute their loading factors. Plot the results.

13. Write a program to implement one of the hashing functions in the text.

9
MEMORY
MANAGEMENT

9.1
The Fragmentation
Problem

In the preceding chapters we have been concerned with choosing data structures to use memory space efficiently. We have not considered the question of where that memory space comes from. We have assumed that the computer system could always supply the necessary memory on demand. Of course this cannot be the case in a system with many users requesting space in a finite memory. The following discussion is an introduction to the problems that confront designers of systems when they must implement procedures for allocation and release of space in memory.

Memory management occurs on at least two levels in a computer system. At the lowest, most fundamental level is the operating system's storage allocation manager. It handles the allocation and release of storage for all the jobs in the system. It keeps track of the free and assigned areas, and provides basic services for all programs. The operating system's storage manager generally works with large blocks of storage, something on the order of thousands of locations.

A higher level of memory management that is part of an application's execution environment is at the user level. The services depend directly on the programming language used in the application's implementation. In languages such as FORTRAN and COBOL, which have static data structures, storage areas are obtained for the program's data and for its object code. When the program performs I/O operations, storage is obtained for input and output buffer areas. Languages such as PL/I and Pascal have dynamic data structures and support recursive procedures, so the execution-time environments of PL/I and Pascal must provide more storage management facilities than that of FORTRAN and COBOL. During execution they must allocate and deallocate data areas as procedures are called and exited. In APL, the user has the capability of dynamically changing a variable's size, type, and shape. Hence the APL storage manager must be able to cope with the possibility of having constantly changing data structures.

Before we go into the management within an application's execution-time environment, we will consider memory management on a system level. We will examine the types of services provided as well as the possible ways of implementing them. Figure 9.1 is a diagram of a system's internal storage area. Each job has a portion of storage allocated for its use. To make our discussion simpler, we imagine that the storage is contiguous. (In a paging operating system, a job's storage may not be contiguous, but the operating system together with the computer's hardware manages that.) As new jobs enter and old jobs finish, we expect to see changes in the allocated space. In Figure 9.1(b) the available space is schematically represented as six units, which is considerable compared with the three to four units required for most jobs in the example. However, it is evident that there is not enough space for a job that requires only three units.

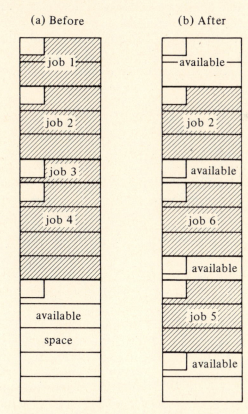

Figure 9.1 Snapshots of system storage area

Figure 9.1 shows something that can and does occur in such systems. A checkerboard pattern of allocated spaces interspersed with available spaces causes loss of usable memory. Even though the total space available might be considerable, it is often too fragmented to use. This type of memory waste is called *external fragmentation*. It results from the fact that we have allocated the exact amount of storage needed by a particular job. When that job finishes, its storage may not be the right size for the next job that comes along.

If the problem of external fragmentation is serious (see Exercise 3), we might consider an alternative scheme to help alleviate the problem. Instead of allowing variable-size blocks for each job, we could require every job to use one or more blocks of fixed size, as shown in Figure 9.2(a). Although our diagram still shows a job occupying contiguous locations, contiguous locations are not necessary. With fixed-sized blocks, job 2 could occupy blocks 2 and 5 instead of 2 and 3. After a period of time, the configuration changes to that shown in Figure 9.2(b). There is still some wasted space in blocks 4 and 6. Job 7 fills block 3, but only part of block 4. Job 5 needs only part of block 6 but gets the whole block. This type of memory waste is called *internal fragmentation*. In our efforts to eliminate external fragmentation, we have succeeded in introducing a different kind of memory waste.

If a system begins with a large block of memory and allocates areas toward one end of memory, the available space will be located at the opposite end of memory. As jobs complete, making their areas available and allowing other

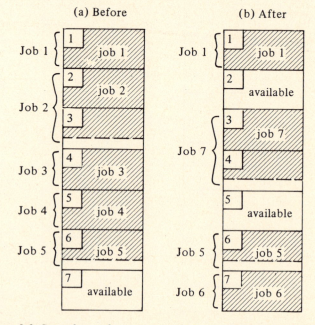

Figure 9.2 Snapshots of system storage area with fixed-size blocks

jobs to enter the system, memory becomes fragmented, as shown in Figure 9.1(b). The available space is unusable because it is separated into small disjoint pieces. A solution to this problem is to move all the allocated space up into a contiguous area and collect the available space at the bottom, or vice versa. This process is called *compaction,* which is a form of garbage collection. *Garbage collection* refers to any technique that makes unused memory areas available for use.

Garbage collection is a treatment for external fragmentation. Because it is expensive and has bad side effects, it should be used only as a last resort. A better solution is prevention. Prevention means better management of memory. Through careful management we can practically eliminate external fragmentation, but internal fragmentation paradoxically becomes a problem.

BEST FIT VERSUS FIRST FIT

Early designers of memory management tried to solve the problem by using this strategy. Scan the entire list of available memory areas and choose the one that best fits the requested size. Allocate the amount requested and return any leftover space to the available list. If the leftover space is "too small" to be of any use, allocate the entire block. This approach is known as the Best Fit strategy.

Scanning the entire list to find the best fit could take an excessive amount of time, so it might be simpler to take the first available block that is large enough to hold the request. The unused portion can be returned to the list of available memory blocks. This approach is the First Fit method. Surprisingly, the First Fit method is often better than Best Fit strategy. For example, suppose that we have available the following list of blocks:

$$1200$$
$$1000$$
$$3000$$

and our list of requests is

$$700$$
$$500$$
$$900$$
$$2200$$

Best Fit strategy allocates storage in the following way:

1. Allocate 700 words from the 1000 word block.
    Result: 1200
    300
    3000

2. Allocate 500 words from the 1200 word block.
    Result: 700
            300
            3000

3. Allocate 900 words from the 3000 word block.
    Result: 700
            300
            2100

4. Cannot allocate storage to the 2200 word block.

However, the First Fit way would first allocate 700 words from the 1200-word block, then 500 words of the remainder. Thus 900 words would be removed from the 1000-word block, and 2200 words from the 3000-word block. This is one instance when the First Fit way works better than Best Fit strategy. There are examples in which the reverse is true (see Exercise 6). Knuth reports simulations with a variety of requests that have shown the First Fit method to be better in terms of number of requests filled per unit time.[1]

With the splitting of available blocks and return of the unused portion to the list, we might expect memory to be subject to severe external fragmentation. What we need is something short of garbage collection to perform this function: If two adjacent blocks (perhaps too small to be useful) become available, combine them into a single block, because the larger block is more useful than the two small blocks. Of course it may be difficult to locate adjacent blocks. In fact, the search mechanism may be very time-consuming. What we need is a systematic method of splitting and coalescing small available blocks into larger ones. Ideally the system would allocate small amounts of memory for small requests, but could still allocate large blocks of memory as necessary. This system is called a *dynamic memory-management system*.

The capability of closely matching allocated space to requested space is the feature that makes a memory-management system a dynamic one. This feature contrasts with the arbitrary allocation of one or more blocks of the same size to every request. In practice, a dynamic memory system may include characteristics of both systems. That is, it may prefer to allocate blocks of certain sizes, but there are enough different sizes to match closely most requests.

The number of distinct block sizes is sometimes called the number of *levels* of the system. All blocks of the same size are considered to be at the same level.

Example   A Five-level Dynamic Memory Management System. Suppose that a memory of 1000 words has blocks of sizes 200, 400, 600, 800, and 1000.

---

[1] Donald E. Knuth, *The Art of Computer Programming*, vol. 1, 2d ed. (Reading, Mass.: Addison-Wesley Publishing Co., 1973), pp. 445–451.

Blocks at level 1 can be coalesced into blocks at level 2; blocks at levels 1 and 2 can combine into a block at level 3, and so on.

Level	Size	Number of possible blocks
5	1000	1
4	800	1
3	600	1
2	400	2
1	200	5

EXERCISES

1. Explain the differences between external and internal fragmentation and the circumstances under which they appear.

2. Develop algorithms to maintain available and allocated lists for the two systems shown in Figures 9.1 and 9.2.

3. Use a pseudorandom-number generator to produce requests for memory space and the length of time that the space will be occupied. Simulate the systems described in this section, allocating and releasing space as requested. If the memory area fills up, keep a queue of requests that cannot be filled until space becomes available. Use the same set of request sizes and hold times for each system and see which one can honor all the requests first. Compute such things as the amount of unallocated memory area, average queue length, and other values that would be useful in comparing the systems.

4. Investigate several garbage-collection schemes.[2]

5. In a multilinked list, each atom contains pointers to one or more atoms. If it is possible for a single atom to be pointed to by several atoms, then an atom that is not referenced can be deleted from the list. Its space can then be freed and reallocated when necessary. To know when to release the space allocated to an atom, a system may use a reference counter. The *reference counter* keeps track of the number of pointers referencing an atom. If it is zero, the atom can be deleted. Discuss advantages and disadvantages of the reference counter. If a node points to itself (like the header of a nil circular list), can its space ever be deallocated?

6. Find an example of available blocks and request sizes that show that sometimes the Best Fit method yields a better allocation than First Fit strategy.

[2] R. Fenichel and J. Yochelson, "A LISP Garbage Collection for Virtual-Memory Computer Systems," *CACM* 12, no. 11 (November 1969):611–612.
Kenneth Knowlton, "A Fast Storage Allocator," *CACM* 8, no. 10 (October 1965):623–625.
H. Schorr and W. Waite, "An Efficient Machine Independent Procedure for Garbage Collection in Various List Structures," *CACM* 10, no. 8 (August 1967):501–506.

7. Write an algorithm to grant a memory request based on the Best Fit method.

8. Write an algorithm to allocate memory based on the First Fit strategy.

9. In view of examples, such as occur in Exercise 6, of what real value is simulation? How should the results of a simulation be interpreted?

10. Perform the following simulation by hand. Assume a memory of 20 words. Each time unit, the system receives a new request for memory. When a request is allocated, it holds its location for five time units. Unfulfilled requests go in a queue, followed by any new requests.

    Use both the First Fit and Best Fit strategies. Take the request sizes from the first 20 digits of π. For example, the first three request sizes are 3, 1, and 4. Adjacent available blocks should be recombined (why?).

9.2
Memory Management Within an Application Environment

Many of the same types of memory-management data structures are used by the operating system's memory manager and an application environment's memory manager. They all contain the same basic components.

1. A list (linked or otherwise) of available space

2. A method for finding storage (Best Fit, First Fit, and so forth)

3. A method for reclaiming storage no longer required

4. A methodology for doing garbage collection

Variations appear in the size of storage allocated, the frequency of requests and releases involved, and the implementation chosen.

An interpreted language such as APL makes heavy use of its storage manager. Almost every APL statement requires some storage for the intermediate results used in a calculation. The basic data structure in APL is the array, and operators can execute on arrays. During execution, variables frequently change size, dimension, or other attributes in addition to value. Thus the possibility exists that new storage may need to be allocated for a variable every time it is assigned a new value. Some APL implementers, after measuring the number of times that the storage manager was required compared with the amount of checking needed to determine whether or not to allocate storage, decided that they were better off allocating new storage to a variable every time it was assigned a new value. The storage previously allocated to the variable can be released. A symbol table keeps track of the variable names, their currently allocated storage areas, and their current attributes.

The constantly changing storage environment of APL tends to leave the storage fragmented with many noncontiguous storage areas, some of which are in use and some of which are available. To keep track of the available space, we may use a linked list that locates the address of each available area and its size. When a request for storage occurs, we can scan the list to find the area that comes closest to matching the requested size. If the list is doubly linked, then one link can define ascending order by size and the other link can define ascending order by address. Thus, when given a request, we can quickly find the smallest block that will fit the request. As a block is freed, it can be placed on the list. The links that order the available spaces by address give us the information to permit the coalescing of two adjacent free blocks into one.

Several key questions to answer as part of this storage-management scheme are: What if the storage request is less than the block of available space? Should the extra space be placed on the available space list? If the storage ever becomes extremely fragmented, should there be total reorganization of storage? To answer the question about the leftover space, there are several possibilities. One is to return the extra space to the available space list if it is larger than a certain percentage of the total. The actual percentage to use is best determined experimentally. It should be one that keeps the storage fragmentation, recombination, and reorganization to a minimum. Another possibility is to have a maximum and a minimum size. Regardless of its percentage of the total, storage smaller than the minimum should not be returned to the available space list. Similarly, storage larger than a certain maximum should probably be returned to the available space, even if it is a small percentage of the total. Reorganization (reassigning variables and moving values around to obtain one contiguous free-space area) may be necessary only if there is enough free space, but no block large enough to fill the current request.

Storage management is an important part of the computer system. Its services are required frequently and its cost must be kept to a minimum. In order to choose an efficient method, one should think about it carefully, analyze it, and test it. Even after a system is implemented, it is possible to measure its effectiveness and efficiency and take appropriate actions if necessary.

EXERCISES

1. What information should be kept in the header of the linked list for the available-space list?

2. Write an algorithm to perform the storage allocation described in this section.

3. Give an example of the available-space list for the APL system. Allocate 50 locations and show the change. Allocate 200 locations, then free the first 50.

4. Where is the available-space linked list kept? Is it part of the space itself or separate from it? Does it matter where the available space is?

9.3
The Buddy System

One memory management system that incorporates the considerations of dynamic memory management is the Buddy system.[3] In the Buddy system, memory is always allocated in sizes that are a power of 2. If a block is much larger than a memory request, it can be split into two blocks, each block size being a power of 2. These blocks are called *buddies*. If both buddies become available, they can be recombined into one larger block. Because blocks are always a power of 2 (2, 4, 8, 16, 32, . . .), it is easy to compute the location of a block's buddy and then determine whether or not a larger block can be formed.

Initially we assume that the entire memory is available and has size $m = 2^n$. This forms an n-level system that offers the following assortment of possible block sizes: $2^n, 2^{n-1}, 2^{n-2}, . . . , 8, 4, 2$. However, at any given moment this does not mean that a block of each type is available or is even in the system. To know which blocks and sizes are available, we use n linked lists whose headers are AVAIL[1], AVAIL[2], . . . , AVAIL[n]. In general, AVAIL[i] points to the location of an available block of size 2^i. If the AVAIL[i] pointer is nil, then no block of size 2^i is free (but could possibly be obtained by splitting a larger block). Thus,

$$\text{AVAIL[1], AVAIL[2], . . . , AVAIL[}n - 1]$$

initially have nil pointers and AVAIL[n] = zero. (We assume that the memory has addresses 0 through 2^{n-1}, as is normally done on computers.) Within each block is a field to specify whether or not the block is free. In this we use

$$\text{FREE}_k = 1 \quad \text{if the block beginning at location } k \text{ is available}$$

$$\text{FREE}_k = 0 \quad \text{if the block beginning at location } k \text{ is not available}$$

Since it is possible that several blocks of the same size may be available, we include a link within each block to point to the next available block of that size. Because there is only one block of size 2^n, that block's link field is always nil. For convenience, the FREE and LINK fields would be part of each block. Since they cannot contain data, we consider them as overhead for the memory-management system. The AVAIL list is also overhead.

As an example, assume that we begin with a memory 32 units long ($32 = 2^5$). Its addresses are 0 through 31, or 00000 through 11111 in binary notation. If the first request is for a block of size 8, the block of size 32 is split into two blocks of size 16. One of these is placed at the head of the

---

[3] Kenneth Knowlton, "A Fast Storage Allocator," *CACM* 8, no. 10 (October 1965):623–625.

AVAIL[4] list, and the other is split into two blocks of size 8. One of these can be allocated, and one can be placed at the head of the AVAIL[3] list. If the next request is for a block of size 4, we find the smallest block that will fill that request. In this case it is a block of size 8. It must be split—one part allocated and one part placed on the AVAIL[2] list. After these operations, our AVAIL list is now

$$AVAIL[1] = nil$$
$$AVAIL[2] = 12$$
$$AVAIL[3] = nil$$
$$AVAIL[4] = 16$$
$$AVAIL[5] = nil$$

and memory appears as below:

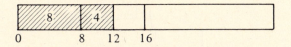

As a block is split, its right half (the part with the large address) remains free and the left half allocated. Now suppose that a request for a block of size 8 is made. To fulfill this request, the block beginning at location 16 is split and memory is now allocated in the following way:

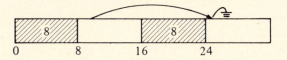

If the block at location 8 becomes free, it can be recombined with its buddy to form another block of size 8. It cannot be combined with the block beginning at location 24, since these blocks are not buddies. However, they can be linked together on the available list. AVAIL[3] will be 8; the block at location 8 will point to the block at location 24, whose link is nil. Memory can be visualized as shown below. The ground symbol (⏚) represents the nil link.

To understand the location of a block's buddy, see Figure 9.3 (p. 216), which shows a possible configuration of memory after many allocations and releases have occurred. The boxes below the addresses in Figure 9.3(a) give the address in binary. The tree structure in Figure 9.3(b) helps us visualize the

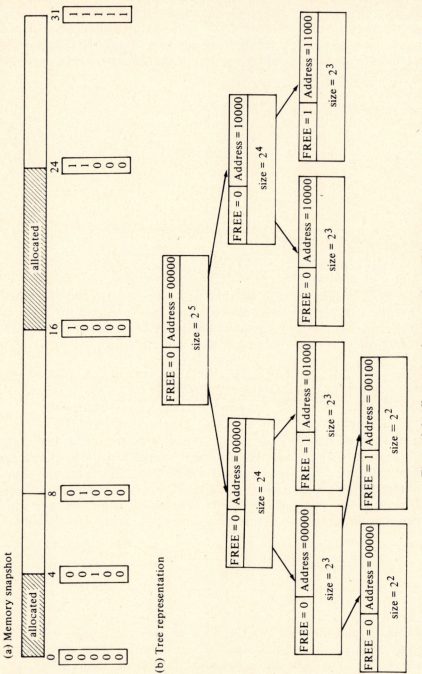

(a) Memory snapshot

(b) Tree representation

Figure 9.3 Allocation of memory by Buddy system

blocks that are buddies. The FREE field shows which blocks are available (FREE = 1) and which are allocated (FREE = 0). We see that there are two available blocks of size 8 and one of size 4. In the tree, nodes preceded by the same parent node are buddies. For example, nodes at addresses 0 and 4 (00000 and 00100) and nodes at addresses 16 and 24 (10000 and 11000) are buddies. Each pair of addresses differs by only a single bit in the ith position, where i is the level number of the block and 2^i is the size of the block. [Here the low-order (rightmost) bit is numbered 0.] A block of size 2^i beginning in location k has a buddy at either location

$$k + 2^i \qquad \text{if the } i\text{th bit of its address is 0}$$

or
$$k - 2^i \qquad \text{if the } i\text{th bit of its address is 1}$$

If the buddy is free, the two buddies can be recombined into a block of size 2^{i+1}. So long as $i + 1 < m$, the new block is not the entire memory area, and more recombinations may be possible.

The Buddy system shows an adaptive feature in its ability to split large blocks and coalesce buddies. However, it may suffer some external fragmentation if free blocks of the same size cannot be recombined because they are not buddies. If requests are not a power of 2, the result is internal fragmentation. Even though we cannot eliminate fragmentation, the Buddy system minimizes waste of memory by adapting to sizes requested as nearly as possible.

The topics and questions presented here are intended to emphasize the use and misuse of data structures. Very often designers or programmers select a particular data structure because they are familiar with it, or perhaps "feel" it is better. What we need is evidence to back up these feelings. One way we can get such evidence is by designing simulation experiments to measure the advantages or disadvantages of the structure in question. (See, for example, Exercise 3 at the end of Section 9.1. For another approach to selection of an efficient memory algorithm, see Section 9.5.)

EXERCISES

1. Is it possible to have external fragmentation with the Buddy system? Discuss ways to measure external and internal fragmentation.

2. Design algorithms to allocate and release space in the Buddy system.

9.4
Fibonacci Memory Management Systems

We now generalize the Buddy system memory-management algorithm. We present the generalized Fibonacci system and demonstrate the use of data

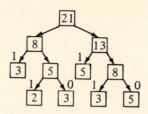

Figure 9.4 A Fibonacci tree for k = 1 in the Fibonacci allocation system

structures in memory allocation during execution of the program. The algorithms given below were obtained from Hinds,[4] Hirschberg, and Knowlton.

The Fibonacci system creates blocks of size F_n, where

$$F_n = F_{n-1} + F_{n-k-1}$$

$$F_0 = 0, \qquad F_1 = F_2, \ldots, F_{k+1} = 1$$

for some arbitrary k. In the case $k = 0$, the Buddy system is realized with blocks of size $F_n = 2^n$. Although the following algorithms are general for any value of k, we will use the case $k = 1$ as a demonstration. The set of blocks 1, 1, 2, 3, 5, 8, 13, 21, . . . would thus replace 1, 2, 4, 8, 16, 32, . . . in the Buddy system discussed earlier. Indeed, we can view the tree of adjacent Fibonacci blocks in a manner similar to the Buddy system tree (see Figure 9.4).

The central problem encountered with generalized Fibonacci systems is not in allocating but in locating adjacent "buddies" for coalescing into free blocks. In Figure 9.4, the subtrees of each node are therefore the adjacent blocks in the Fibonacci system. For example, at the lowest level, 2 and 3 coalesce to yield 5, 5 and 3 yield 8, and 5 and 8 yield 13.

A method that ensures that left and right buddies are matched uses a "left buddy counter" that is zero for all terminal nodes with a left buddy, and nonzero for all terminal nodes with a right buddy. In Figure 9.4 we see that only two nodes have a left buddy. We can extend this notion so that it will be useful during "split" and "free" operations. We maintain the counter as follows:

1. The root node (the largest block) is assigned a LEFT-BUDDY-COUNT of zero.

2. At each split, assign the LEFT-BUDDY-COUNT of the LEFT and RIGHT BLOCKS recursively:

LEFT-BUDDY-COUNT [RIGHT-BLOCK] = 0

[4] James A. Hinds, "A Design for the Buddy System with Arbitrary Sequences of Block Sizes," Technical Report No. 74, State University of New York at Buffalo, Computer Science Department, 1974.

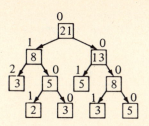

Figure 9.5 The left-buddy-count of the tree of Figure 9.4

LEFT-BUDDY-COUNT [LEFT-BLOCK] =

LEFT-BUDDY-COUNT [PARENT] + 1

We can now draw the tree for Figure 9.4, applying the rule above in order to obtain the LEFT-BUDDY-COUNTS prescribed (see Figure 9.5).

The two buddies that have a LEFT-BUDDY-COUNT of zero and nonzero are candidates for coalescing. If they both become free, they are coalesced, and the LEFT-BUDDY-COUNT of the coalesced pair is one greater than the count of the left subtree.

EXERCISES

1. Design a linked-list AVAIL structure for the Fibonacci allocation system and give an ALLOCATE and FREE algorithm.

2. Generate the first 10 block sizes from the $k = 2$ Fibonacci system. Assume that $F_0 = 0$, $F_1 = 1$, $F_2 = 1$, $F_3 = 1$.

3. Show that the formula $F_n = C\alpha^n$ satisfies the Fibonacci recurrence formula for any value of k (specified beforehand). The integers in the sequence are given by this formula by rounding to the nearest integer. (*Hint:* α is the root of $X^{k+1} - X^k - 1 = 0$, and C is obtained by noting that $F_0 = 0$, $F_1 = \ldots$, $F_{k+1} = 1$.)

9.5
A Method for Optimizing Memory Management

You should read this section only if you have the mathematical background to follow it. Although it contains an analytical method of block size selection, it is not essential for understanding memory management. However, it is a fascinating application that will interest the Fibonacci number enthusiast.

We can design a memory-management scheme based on the kth Fibonacci sequence and modeled after the Buddy system. Initially, memory is the size of an appropriate Fibonacci number, and requests for smaller pieces of memory are serviced by using the formula for F_n to split and reassemble blocks. Does

Table 9.1   Generalized Fibonacci Sequences Giving Block Sizes 1
Through 250 (approx.)

i Level	F_i				
	$k = 0$	$k = 1$	$k = 2$	$k = 3$	$k = 4$
1	1	1	1	1	1
2	2	1	1	1	1
3	4	2	1	1	1
4	8	3	2	1	1
5	16	5	3	2	1
6	32	8	4	3	2
7	64	13	6	4	3
8	128	21	9	5	4
9	256	34	13	7	5
10		55	19	10	6
11		89	28	14	8
12		144	41	19	11
13		233	60	26	15
14			88	36	20
15			129	50	26
16			189	69	34
17			277	95	45
18				131	60
19				181	80
20				250	106
21					140
22					185
23					245

this improve utilization of memory? Table 9.1 shows that there is a greater variety of block sizes as k increases, but that system overhead increases.

For the moment we will disregard the overhead and examine the cost due to internal fragmentation. Let $\{F_i\}$ for i from 0 to n be the collection of block sizes, with $F_0 = 0$ and $F_n = m$, the memory size. If the system services a request for a certain number x of memory locations, it will allocate a block of size F_i, where $F_{i-1} < x \leqslant F_i$. The waste involved is $F_i - x$.

The requests for memory space are always for an integral number of locations. Let us assume for convenience that the request sizes are given by a continuous probability function $pdf(x)$. The expected average waste per request $\bar{w}$ is then given by Hirschberg (see References):

$$\bar{w} = \sum_{i=1}^{n} \int_{F_{i-1}}^{F_i} (F_i - x)pdf(x)\, dx$$

Rewriting this, we obtain

$$\bar{w} = m - \bar{x} - \sum_{i=1}^{n} (d_i)cdf(F_{i-1})$$

where m = maximum memory size

$$\bar{x} = \int_0^m x \, pdf(x) \, dx = \text{average request size}$$

n = number of distinct block sizes

$d_i = F_i - F_{i-1}$

$$cdf(z) = \int_0^z pdf(x) \, dx = \text{cumulative distribution function}$$

The objective of memory management is to minimize $\bar{w}$ for a given $pdf(x)$. If we restrict our attention to Fibonacci-type systems, we can gain some additional insight into minimizing $\bar{w}$.

To solve for F_n in closed form, we must investigate the characteristic polynomial $x^{k+1} - x^k - 1 = 0$ of the kth Fibonacci sequence. The polynomial (for fixed k) is known to have $k + 1$ distinct roots, which yield a closed-form expression for the nth Fibonacci number. Note that $f(x) = x^{k+1} - x^k - 1$ has a real root between 1 and 2, since $f(1)$ is negative and $f(2)$ is positive. By Descartes' rule of signs, this is the only positive root, which will be denoted by α_1. Thus $1 < \alpha_1 \leq 2$ ($\alpha_1 = 2$ if $k = 0$). Let the other roots of $f(x)$ be α_2, $\alpha_3, \ldots, \alpha_{k+1}$. It is easy to establish that α_1 is the root of largest modulus and, in fact, $|\alpha_i| < 1$ for $i = 2, 3, \ldots, k + 1$.

Evidently any sequence of numbers $\{u_i\}$ satisfying $u_i = c_1\alpha_1^i + c_2\alpha_2^i + \cdots + c_{k+1}\alpha_{k+1}^i$ will be a kth Fibonacci sequence for F_n. Specifying the initial $k + 1$ terms of the sequence determines the constants $c_1, \ldots, c_{k+1}$; or specifying the constants determines the sequence. For the particular sequence $\{F_i\}$ in Table 9.1, we can write $F_i = c_1\alpha_1^i + \cdots + c_{k+1}\alpha_{k+1}^i$. Since $|\alpha_i| < 1$ for $i = 2, 3, \ldots, k + 1$, it follows that for sufficiently large i,

$$F_i \cong c_1\alpha_1^i$$

Table 9.2 gives some approximate values of c_1 and α_1. The initial segments in Table 9.1 can be obtained from the formula $F_i = c_1\alpha_1^i$ (rounded to the nearest integer).

Consider now the value of α_1 for different values of k. Let this root be denoted by $\alpha(k)$. We have observed that $1 < \alpha(k) \leq 2$. We see that

$$\alpha(k) = 1 + \left[\frac{1}{\alpha(k)}\right]^k$$

and it follows that $\alpha(k + 1) < \alpha(k)$ for every k. In fact, $\lim_{k \to \infty} \alpha(k) = 1$.

Table 9.2   Generators for Fibonacci Sequences

	$k = 0$	$k = 1$	$k = 2$	$k = 3$	$k = 4$
$c(k)$	1	.44721	.41724	.39663	.38119
$\alpha(k)$	2	1.61803	1.46557	1.38028	1.32472

Let us apply the preceding observations to a particular example, in which the distribution of request sizes is given by the uniform distribution $pdf(x) = 1/m$. Then $cdf(x) = x/m$, and $\bar{x} = m/2$. Let k be arbitrary but fixed. We write $\alpha(k) = \alpha$ and $c(k) = c$, so that $F_i \cong c\alpha^i$ and $F_n = c\alpha^n = m$. Then,

$$\bar{w} = m - \bar{x} - \sum_{i=1}^{n} (d_i) cdf(F_{i-1})$$

$$= m - \frac{m}{2} - c\left(1 - \frac{1}{\alpha}\right) \sum_{i=1}^{n} \alpha^i \, cdf(c\alpha^{i-1}) = \frac{m}{2} - c\left(1 - \frac{1}{\alpha}\right) \sum_{i=1}^{n} \alpha^i \frac{c\alpha^i}{\alpha^m}$$

If we assume that $m \gg 1$, then

$$\alpha^2 m^2 - 1 \cong \alpha^2 m^2 \qquad \text{and} \qquad \bar{w} \cong \frac{m}{2} - \frac{\cdot m}{\alpha + 1}$$

Thus $\bar{w}$ can be made as small as desired by increasing k, since α approaches 1 as k increases.

Intuitively, this is to be expected, since for any finite memory size m, if $k > m$, the kth Fibonacci sequence contains all the integers from 1 through m, and $\bar{w}$ should be zero. However, this leads to extreme overhead in memory management and places unreasonable demands on the search mechanism for allocation and release of areas in memory.

The waste function $\bar{w}$ measures only the cost of internal fragmentation. Let us assume that the overhead associated with a memory system is given by a function of n, the number of distinct block sizes. A more complete cost function is then

$$w = m - \bar{x} - \sum_{i=1}^{n} (d_i) cdf(F_{i-1}) + f(n)$$

This raises the possibility of optimizing the collection $\{F_i\}_{i=0}^{n}$ by considering the equations

$$\frac{\partial w}{\partial F_j} = 0 \qquad \text{for } j = 1, 2, \ldots, n-1$$

and the boundary conditions $F_0 = 0$, $F_n = m$. The solution is given by the "state equation"

$$F_{j+1} = F_j + \frac{cdf(F_j) - cdf(F_{j-1})}{pdf(F_j)}$$

Continuing with the simple example of requests from a uniform distribution, let us assume conveniently that $f(n) = \beta \cdot n$, where $\beta > 0$ is a constant. We obtain

$$F_{j+1} = F_j + \frac{\dfrac{F_j}{m} - \dfrac{F_{j-1}}{m}}{\dfrac{1}{m}} = 2F_j - F_{j-1}$$

The difference equation is not of the Fibonacci type, but does have a closed-form solution:

$$F_j = \frac{m}{n}j \qquad \text{for } j = 0, 1, \ldots, n$$

It is therefore possible to optimize the collection $\{F_j\}$, which minimizes w, provided that we know the nature of $f(n)$.

Unfortunately, other request distributions and other functions $f(n)$ do not lead to such nice solutions. Indeed, the difference equations resulting from the state equation are, in general, extremely difficult to solve analytically. For certain $pdf(x)$, however, solutions are of considerable importance to designers of computer systems. And where closed-form solutions of the difference equations are not feasible, it is still important to apply numerical techniques to these problems.

EXERCISES

1. Write a computer program to compute F_n for any n and any k.

2. Write a computer program to compute $\bar{w}$ for any value of n, any $pdf(x)$, and any F_i.

3. Devise an algorithm to ALLOCATE and FREE the blocks that are optimal for $pdf(x) = 1/m$. Assume a fixed value of n.

4. Assume that

$$pdf(x) = \frac{1}{\sigma m} e^{-x/m} \qquad \text{for } 0 \leq x \leq m \qquad \text{and} \qquad \sigma = \int_0^m \frac{1}{m} e^{-x/m} \, dx$$

Calculate the optimal block sizes for this system when $n = 10$.

10
ADVANCED APPLICATIONS

The following sections cut across three phases of computer programming: preparation, translation, and execution. We wish to demonstrate how data structures permeate these three phases, and to integrate the ideas from the material covered in previous chapters.

We follow an evolutionary development of the growth of data structures and table operations required in computer programming. In this way, you can see the relationships among the many parts already described. In particular, the parts are bound together by our need to add, delete, and change information within computer memory.

PROGRAM PREPARATION

The design of a computer program begins with the definition of a problem and the organization of the tasks to be performed. We use a flowchart, flow diagram, or English-like algorithm to describe the tasks identified by our analysis. It shows the manipulation of the data. The operations used in manipulation determine what structure is most efficient for storage of data. Therefore an awareness of data-structure techniques is necessary at the earliest stage of computer programming: program preparation.

We demonstrate three data structures imposed by the preparation of programs for three diverse applications. In the analysis, we assume no considerations of programming language (this may be unrealistic). The applications are numerical calculations on sparse arrays, updating operations on a structure for display graphics, and text editing of source statements. The applications may be studied in any order, since they are independent examples.

10.1
Sparse Arrays

We will study a common problem in numerical calculations: efficient storage of sparse arrays. Suppose that a two-dimensional array of 100 by 100 elements is needed to do a certain calculation. The array space for implementing a dense list would require 10,000 memory locations. Suppose further that all but three diagonals of the array contain zero. An array containing many zero elements and few nonzero elements is termed a *sparse array*. We define a tridiagonal array with elements $a_{i,j}$, where $i = 1, 2, \ldots, n$ and $j = 1, 2, \ldots, n$ as in the form shown below.

$$A = \begin{bmatrix} a_{11} & a_{12} & 0 & \cdots & & 0 \\ a_{21} & a_{22} & a_{23} & \cdots & & 0 \\ 0 & a_{32} & a_{33} & & & \\ 0 & & & \cdot & & \\ \cdot & & & & \cdot & \\ \cdot & & & & & \\ \cdot & & & & a_{n-1,n-1} & a_{n-1,n} \\ 0 & & & & a_{n,n-1} & a_{n,n} \end{bmatrix}$$

In the array under discussion, $(98 \times 3) + 4 = 298$ entries are nonzero, while $10{,}000 - 298 = 9{,}702$ entries are zero in the 100×100 array. Much space is being wasted in storing zeros. Is there a data structure that will save memory in this case? Let us use a 3×3 array example to show how we could save space in the 100×100 array problem.

Figures 10.1 and 10.2 demonstrate a multilinked list structure for storage of a 3×3 array. In Figure 10.2 only two atoms are nonzero. Hence only two atoms are included in the list. The variables named $a_{1,1}$, $a_{2,1}$, and so forth represent the values in the array.

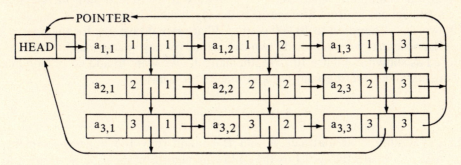

Figure 10.1 Data structure for a 3×3 matrix

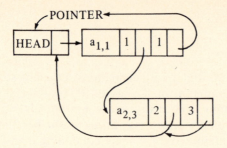

Seven words of memory are saved by using a linked list instead of a dense list for a sparse matrix.

Figure 10.2 Data structure for a 3 × 3 matrix with [1, 1] and [2, 3] elements nonzero

The row pointers of Figure 10.1 establish a ring structure for rows, while column pointers encircle vertical columns. Note that additional overhead is needed to label each atom with appropriate subscripts. These are needed to identify which element of the matrix is nonzero. The $a_{i,j}$ entries would contain the actual value stored in the i,j position of the matrix.

If we assume that each atom of data in Figure 10.1 requires two words of space [one for $a_{i,j}$ and one for (i,j) plus pointers], then the density of an $n \times n$ array is as follows:

$$d = \frac{n^2 \text{ words of data}}{2n^2 \text{ words of data structure}} = \frac{1}{2}$$

On the other hand, the density of a dense list is 1.0. The break-even point for "sparseness" is 50%. In the case of a 100×100 array, if more than 5000 entries are nonzero, using a dense list would save more space than using a multilinked list that records only the nonzero entries. In general, if the density of the sparse-array multilinked list is d, then $d \cdot n^2$ nonzero entries is the break-even point.

In our example, 298 nonzero elements are to be stored and manipulated as a tridiagonal matrix. Storage as a doubly linked list requires $2 \times 298 = 596$ words, as opposed to 10,000 words! What price do we pay for this saving?

Suppose that we wish to add two tridiagonal (or any sparse matrix) matrices stored as shown in Figure 10.1. Mathematically we want to add corresponding elements and store the sum back in one of the arrays:

$$a_{i,j} \leftarrow a_{i,j} + b_{i,j} \qquad \text{for all } i,j$$

Let us assume that matrices A and B contain only the nonzero elements $a_{i,j}$ and $b_{i,j}$, respectively. Thus the following algorithm must be performed to accomplish the addition indicated above.

Algorithm   ADD_SPARSE

1. Comments: This algorithm adds the sparse matrices A and B. The result is stored
                in A. The algorithm assumes that neither matrix is empty. The op-
                eration of "Get next element" follows the row and column pointers
                and tries to pick the next element on the same row. When one row
                has been traversed, the selection proceeds to the next row.

    Input: A   Matrix
                HEADA   Header for matrix A; locates first nonzero element
                B   Matrix
                HEADB   Header for matrix B; locates first nonzero element
    Output: A   Matrix
    Variables: IA, JA   Subscripts for matrix A
                IB, JB   Subscripts for matrix B
                PA, PB   Pointers to an element in matrix A and B, respectively

2. Locate first nonzero element in matrices A and B:
    PA ← HEADA; PB ← HEADB

3. Find subscripts of first nonzero A and B element:
    IA ← I subscript of A element pointed to by PA
    JA ← J subscript of A element pointed to by PA
    IB ← I subscript of B element pointed to by PB
    JB ← J subscript of B element pointed to by PB

4. Repeat the following until one of the matrices has had all its elements processed:
    If IA < IB, then
       Get next element of A
    Else
       If IA = IB, then
          If JA < JB, then
             Get next element of A
          Else
             If JA = JB, then
                A[IA,JA] ← A[IA,JA] + B[IB,JB]; Get next elements of A and B
             Else
                Insert B[IB,JB] in array A[IB,JB]; Get next element of B
       Else
          Insert B[IB,JB] in array A [IB,JB]; Get next element of B

5. If array A has had all its elements processed, then
    Insert the rest of array B into array A

6. End ADD_SPARSE

Operations on sparse matrices stored as linked lists are more complex and
time-consuming. Therefore we should use dense structures, unless storage is
limited or the application warrants storage-minimizing usage.

In preparing a computer program for processing a sparse array, we must
choose the data structure and then implement a search algorithm analogous

to the one used in ADD_SPARSE (unless the programming language incorporates the search mechanism). The selection of structure is made during analysis and design of the computer program, that is, during program preparation.

EXERCISES

1.   Devise an algorithm for locating the (i,j)th element of a sparse matrix stored in a multilinked list.

2.   Use the results of Exercise 1 to write an algorithm to multiply two sparse matrices.

3.   Design a data structure for a sparse vector, and then for a three-dimensional array. What can you say about the density of each and their storage efficiency breakeven point? Devise a hashing algorithm that transforms the subscript values of (i,j) into a table address corresponding to the $a_{i,j}$th element of a sparse array, A. Discuss this approach as compared with the linked-list approach.

10.2
A Display
Graphics Structure

The application of linked lists to display graphics data demonstrates again the usefulness of linked lists and the need to analyze a problem fully before implementing it. Two-dimensional input/output, employing a televisionlike device for output and a light-pen device for input, is a good means for displaying large volumes of data to people. Such large volumes of data require sophisticated structuring to make the data presentable or viewable.

Either a refresh-type display (the picture must be "repainted" 30 or 40 times a second to keep it from fading) or a storage-type display (the picture is painted once and remains until "erased") is used to display data stored in structures appropriate to the device.

We will discuss data structures that are especially suitable for refresh-type cathode-ray tube (CRT) terminals. Storage-type CRT terminals are also used in a slightly different manner, but the data structuring is quite similar.

Before one can draw a single point or line on the face of a CRT, one must know a convention for coding output data. These conventions are called *plotting modes*.

Modes are established by the CRT hardware. The modes of Figure 10.3 (p. 230) show some possibilities for plotting points, vectors, and lines. In all cases, a grid of Cartesian coordinate points is assumed to be transparently superimposed on the face of the CRT. Each point on the grid represents a point that may be referenced by a plotting mode word.

The basic modes that use this invisible grid are:

POINT: Plot a point at location (X, Y).
VECTOR: Draw a line from the current point (X, Y) (last point
              drawn) to the point (X + ΔX, Y + ΔY).
LINE: Draw a line from (X_1, Y_1) to (X_2, Y_2).
INCREMENT: Draw short line segments in steps of size R, where R
              = resolution (size of a grid "square").

The escape bit is used to switch from one mode to another. Normally the CRT plotting hardware assumes a mode setting given at the beginning of a sequence of plots.

The relative merits of different modes depend on the application and attempts to conserve bits. The best mode may be a combination of all the modes given. For example, 150 lines may be drawn in INCREMENT mode and then 28 lines are drawn in the LINE mode by use of the escape bit to signal a change.

The data structure is actually built as a bit string, usually in multiples of bytes or fractions of words to conserve space. The CRT hardware prepares a point for plotting by storing a plotter word in separate *display registers,* called the *x* and *y* axis registers. The *x*-display and *y*-display registers are separately converted from digital into analog signals, which position and fire an electron beam.

An example, demonstrating a data structure for line mode, is given for the "graph" of Figure 10.4 (p. 231). This picture is composed of components defined in the data structure using BACKWARD pointers (see Figure 10.5). The significant point here is that the BACKWARD pointers are used to group together components of the picture.

Component 1 consists of lines 1 through 6. These lines are represented in the data structure of Figure 10.5 by atoms 1 through 6. The BACKWARD pointer of atom 6 points back to the HEADER, thus signifying a component. Component 2 consists of lines 7 through 9 and is represented by atoms 7 through 9 in Figure 10.5. The BACKWARD pointer of atom 9 groups together all atoms in component 2. In this way we can manipulate either an atom or a component of data.

The CRT display algorithm paints a picture from the data structure of Figure 10.5 by scanning the FORWARD pointers. Each atom is accessed in turn and a value is copied into the *x, y* display registers.

The data provided in Figure 10.5 are not detailed enough to specify how lines are to be actually drawn (painted), but serve only to demonstrate the overall structure of data atoms. As an example, let us propose a format for the atoms of Figure 10.5. In Figure 10.6 we see that each atom in the data structure is composed of fields.

The algorithm for drawing the picture of Figure 10.4 operates on the data structure of Figure 10.5 containing atoms structured as shown in Figure 10.6. By tracing the FORWARD pointers, the algorithm locates *x, y* values for the

POINT-PLOTTING

0	1	10	11	20	21
I	X		Y		E

I	Intensity (on or off)
X	10-bit X-coordinate ⎫ Any point on a
Y	10-bit Y-coordinate ⎬ 1024×1024 grid
E	Escape bit

VECTOR MODE

0	1	2	8	9	10	16	17
I	S	ΔX		S	ΔY		E

I	Intensity
S	Sign bits
ΔX	Change in horizontal position ⎫ Updated from
ΔY	Change in vertical position ⎬ previous point

LINE MODE

0	1	10	11	20	21	30	31	40	41
I	X_1		Y_1		X_2		Y_2		E

I, E	As before
X_1, Y_1	Starting point ⎫ Line drawn
X_2, Y_2	Stopping point ⎬ between

INCREMENT MODE

0	1	2	3	5	6	7	8	10	11
I	N_1		D_1		N_2		D_2	E	

I, E	As before
N_1, N_2	Number of steps in direction D_1, D_2
D_1, D_2	Direction of steps (in $45°$ increments)

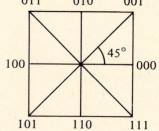

Figure 10.3 Data structure of CRT plotting modes

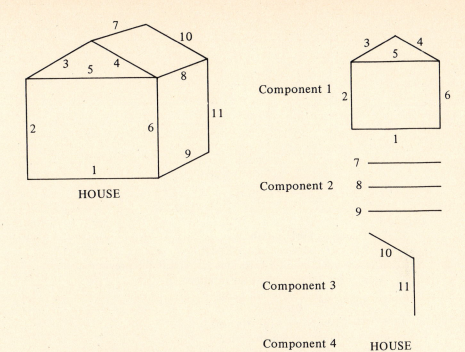

Figure 10.4 Graph of a house to be displayed. Each component is on a different plane in the orthograhic projection.

x, y display registers. If the display code (item 5) is set to 1, the electron beam is directed to paint the given line.

The last component spells "HOUSE" by furnishing the CRT hardware with codes for "H", "O", "U", "S", and "E." If special hardware exists for character generation, these characters are painted in a manner similar to the other components in the picture.

Note that the data structure helps us select lines, components, and pictures as a simple unit of information. This demonstrates how careful design of a data structure can help ease the programmer's burden.

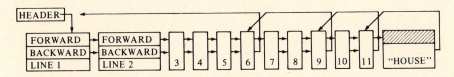

Figure 10.5 Simplified representation of data structure for the graphic display of Figure 10.4

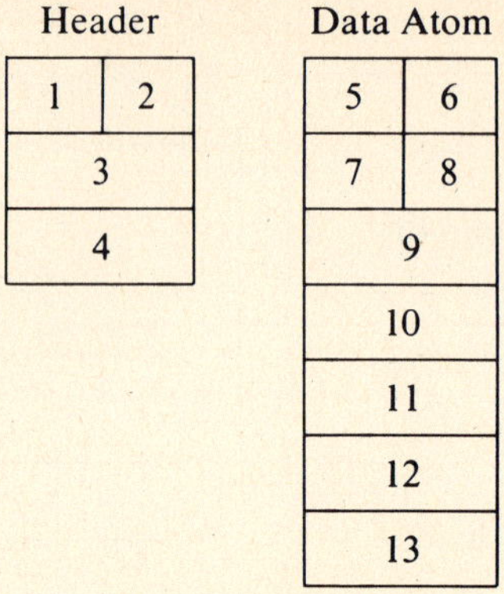

Header | Data Atom

Header	
1	2
3	
4	

Data Atom	
5	6
7	8
9	
10	
11	
12	
13	

Legend
1. User code (console number)
2. FORWARD pointer (next HEADER for another picture)
3. BACKWARD pointer (previous HEADER for previous picture)
4. Atom size (number of words in atom)
5. Display code $= \begin{cases} 0 & \text{do not paint} \\ 1 & \text{paint} \end{cases}$
6. FORWARD pointer (next atom)
7. Plane number (which component)
8. BACKWARD (for grouping components)

9–12. X_1, Y_1, X_2, Y_2 for LINE mode (other modes could be used)
13. Mode switch (set Escape or not)

Figure 10.6 Organization of data for the structure of Figure 10.5

EXERCISES

1. Give a vector mode sequence of plotting words for the "house" of Figure 10.4. Make up your own scaling factor and resolution.

2. How might we expand the INTENSITY field of the plotting-mode words to include three colors?

3. Suggest several plotting modes for a polar coordinate graphics CRT.

4. How might a polar coordinate CRT be used to display complex plane data (imaginary as well as real-valued)?

10.3
Text Editing

Once the proper data structure has been selected for a given problem, a program must be written and debugged. An aid to constructing programs using on-line systems is the text editor. A text editor is a program that helps us type statements into a file. The file can contain the paragraphs, titles, and other parts of a report to be formatted and printed by a word-processing system. Or it can contain programming-language statements that will be processed by a language translator. (We use the term "translator" to include compilers, interpreters, and symbolic assemblers.)

The variety of text editors that currently exist are usually capable of manipulating text in many ways. We wish to discuss the data structuring necessary to do three basic text-editing operations: insertion, deletion, and replacement. These three operations are basic to all text editors, and they also motivate designers of editors to construct elaborate data structures. Suppose that we wish to type in the following segment of source statement program for processing by a PL/I compiler.

> .
> .
> .
>
> *DECLARE* A(10) *FIXED*;
> *GET LIST* (A);
> A(1) = (A(1) − A(2) / A(1) * 5) + 3;
>
> .
> .
> .

This segment shows that A is an array of ten integer (FIXED) numbers. The values of A are input by the GET LIST command, and then A(1) is computed as

$$A(1) - \frac{A(2) \cdot 5}{A(1)} + 3$$

This is input by the editor as follows. Each statement is numbered in ascending order and input, for example, through a keyboard terminal. Errors are corrected by either deleting a line or replacing parts of a line. The final version might appear as below.

> 10 *DECLARE* A(10) *FIXED*;
> 20 *GET LIST* (A);
> 30 A(1) = (A(1) − A(2) / A(1) * 5) + 3;

This appears innocent at first glance, but we see on closer examination that the dynamic nature of this process causes problems for memory organization. Suppose, for example, that we delete statement 20. The adjacent statements, 10 and 30, could be moved to contiguous locations or some other structure could be used to order the statements. Suppose instead that we wish to increase the length of statement 20 by inserting a comment. How is space expanded? A data structure that allows interaction with the characters of each line is needed. We propose here only one of many possible ways to organize text. (Other methods are discussed in Section 2.3.)

Figure 10.7(a) shows how memory can be utilized in text editing. The HEADER is a dense list (often called a dope vector) containing a FLAG, STATEMENT NUMBER, and ADDRESS field. The FLAG field indicates whether the statement has been inserted or deleted.

$$\text{FLAG} = \begin{cases} 0 \text{ inserted statement} \\ 1 \text{ deleted or empty} \end{cases}$$

The STATEMENT NUMBER corresponds to the number used in entering the statement. This number is used to identify the entire statement; any string

(a) Layout of text editor memory

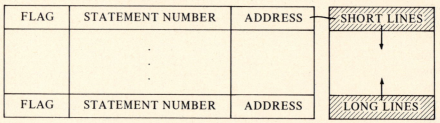

(b) Storage of text given as an example

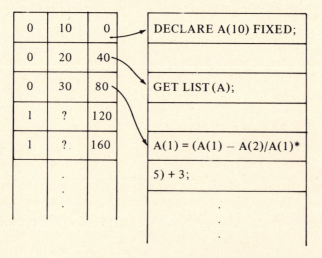

Figure 10.7 A possible data structure for text editing

operations (see Chapter 2) can refer to this number. The ADDRESS is an index into the WORKING SPACE.

WORKING SPACE is divided into two levels: SHORT LINES are text lines of 40 or fewer characters and LONG LINES are text lines of 41 to 80 characters. This method causes memory loss resulting from internal fragmentation of the WORKING SPACE, but it has the advantage of allowing large numbers of updates (increase or decrease of the length of the statement) without the need to move large numbers of characters.

More economical usage of memory is possible using a multilevel system (see Chapter 9 on memory management). Note that the SHORT LINE space grows from the "top" of our SPACE, while the LONG LINE space grows from the "bottom." When these two areas meet, the WORKING SPACE is said to have overflowed.

When overflow occurs, we must compact the two areas by moving the allocated space toward each end and leaving the available space in the middle. The FLAG field is used for this purpose. FLAG = 1 indicates that the area pointed to by the ADDRESS field is available for use. If, however, all FLAGS equal zero, the system is saturated and garbage collection is of no use.

Suppose that we wish to insert a statement numbered 15 into the structure of Figure 10.7(b). We have two choices: We can move everything below 10 "down" to make room for insertion of statement 15, or we can place statement 15 in the first available location shown by the dense HEADER list. If we do the first (move), we incur overhead in moving. If we do the second, we incur overhead in searching for statement 15 later.

This data structure must be flexible enough to allow renumbering statements, creating ordered lists, and searching for substrings. We have outlined merely one approach to this problem, called the *dope-vector* method. Other approaches may use hashing, multiple linked lists, or techniques to compress the text. In each case, the structure is selected because it facilitates the operations to be performed.

EXERCISES

1. Compute the expected internal fragmentation (use layout of Figure 10.7) given that the probability distribution for statement length is $e^{-x/5}/5; x > 0$.

2. Give INSERT, DELETE, and REPLACE algorithms for the data structure suggested in Figure 10.7. The REPLACE operator must find a substring s_0 and replace it with string s_1. Keep in mind that the replacement string may be null.

3. Discuss limitations or drawbacks to the method of Figure 10.7.

4. Design a text-editor system based on hashing instead of the dope-vector method presented here.

5. Write a computer program to perform INSERT and DELETE operations on a data structure. What problems did you encounter?

10.4
Symbol
Tables

Let us use the PL/I program segment of Section 10.3 to illustrate the ideas presented here. We assume that these statements are part of a program entered into a text-editor file. This file is processed by a translator that reads source statements and generates object statements for a particular computer. The recognition of source language statements of a high-level programming language is our main concern.

$$DECLARE\ A(10)\ FIXED;$$
$$GET\ LIST\ (A);$$
$$A(1) = (A(1) - A(2) / A(1) * 5) + 3;$$
$$\cdot$$
$$\cdot$$
$$\cdot$$

These source statements will be used to demonstrate symbol tables for use in translation, templates, or dope-vectoring arrays, and for parsing arithmetic expressions by a push-down stack.

Each variable name in a program is associated with a list of attributes. The attributes include the variable's address, its type, and sometimes an initial value (and other attributes not discussed here). The attribute list is built during translation and then referenced by the compiler.

Figure 10.8 illustrates the structure of such a symbol table. The table is accessed frequently during compilation, so we need a method that allows fast access to the entries, even when the table is being built. Hashing offers us those advantages. The algorithm for searching the symbol table is as follows. A similar algorithm is used to insert entries in the symbol table.

INDEX	NAME	TYPE	ADDRESS
0			
1			
2			
		$\cdot$ $\cdot$ $\cdot$	
(N − 1)			

Figure 10.8 Format for a symbol table

Algorithm   SYM_HASH

1. Comments: Hashing algorithm for symbol table. It assumes that no deletions occur in the symbol table.

   Input: NAME   A variable name to be hashed

   N   Number of entries in symbol table

   Output: INDEX   The hash value of the variable NAME; an index to the table entry belonging to name; 'nil' if name is not in table

   Variables: R   Remainder; result of a division hashing algorithm

   Q   Quotient; used to locate entry if a collision occurs

   FOLDED   One word or location into which NAME can be folded

2. Fold-shift NAME so that it fits into one computer word FOLDED.

3. Perform a division hash:
   R ← FOLDED mod N
   Q ← FOLDED / N

4. Ensure that Q, used to resolve collisions in step 7, is nonzero:
   If Q mod N = 0, then
       Q ← 1

5. Set INDEX ← R

6. If TABLE[INDEX] contains NAME, then
       EXIT SYM_HASH
   Else
       If TABLE[INDEX] is empty, then
           INDEX ← nil
           EXIT SYM_HASH

7. Repeat the following a maximum of N times:
   Add Q to INDEX and continue search:
       INDEX ← (INDEX + Q) mod N
       If TABLE[INDEX] contains NAME, then
           EXIT SYM_HASH
       Else
           If TABLE[INDEX] is empty, then
               INDEX ← nil
               EXIT SYM_HASH

8. N probes have been unsuccessful:
   Output 'Name not in Table'

9. End SYM_HASH

Let us perform the preceding hashing algorithm for the array 'A' and the literals '5' and '3' as given in the sample program segment. The EBCDIC coding for these is (in hexadecimal);

Character	EBCDIC
'A'	$C1_{16}$
'5'	$F5_{16}$
'3'	$F3_{16}$

We will assume a word with 4 bits of precision. Therefore we must shift and fold the two hex digits as follows.

Character	Fold-shift
'A'	$C + 1 = D_{16}$
'5'	$F + 5 = 4_{16}$
'3'	$F + 3 = 2_{16}$

Since we encounter 'A' first in the program, then '5', and finally '3', we will perform the hash on these names in the order: 'A', '5', '3'. The algorithm produces the following results, assuming that $N = 11$.

Variable	Calculation	
R	$D_{16} \bmod 11 = 2_{16}$	
Q	$D_{16}/11 \quad = 1$	For 'A'
INDEX	$= 2_{16}$	

Variable	Calculation	
R	$2 \bmod 11 = 2$	
Q	$2/11 = 0$; set to 1	For '3'
INDEX	$2 + Q = 3$	

Note that the hash value of INDEX in the case for '3' collides with the INDEX for 'A'. The collision is resolved by addition of Q, as indicated in step 4 of the algorithm.

Variable	Calculation
R	4 mod 11 = 4
Q	4/11 = 0; set to 1
INDEX	= 4

For '5'

The final result of storing these names in the symbol table appears in Figure 10.9. Note that we have also filled in the other information to demonstrate a possible outcome for a compiler. We have used the character form of information to assist in reading the names. The address field represents locations where additional information about the array A and the constants 3 and 5 are stored.

It would appear that the information in Figure 10.9 has solved our problem of translating symbolic names in a high-level language into attributes, such as type and address. Note, however, that 'A' is an array. We have not supplied information about the length of A or the number of subscripts for A. For this reason, an additional descriptor called the *dope vector* is used (see also Section 10.5).

EXERCISES

1. Write a computer program to perform the SYM_HASH symbol-table hashing algorithm given in the text.

INDEX	NAME	TYPE	ADDRESS
0			
1			
2	'A'	'ARRAY'	0A54
3	'3'	'CONSTANT'	0BC0
4	'5'	'CONSTANT'	0BC4
. . .		. . .	
10			

Figure 10.9 Result of hashing into symbol table

2. Use the program of Exercise 1 to study the performance of hashing as the loading factor is changed from $\alpha = 0.5$ to $\alpha = 0.9$. Compute the average number of comparisons expected when inserting a symbol.

3. Construct a symbol table based on a binary search tree rather than a hash code table. Show how the tree would appear for the data in Figure 10.9.

4. Create a symbol table for names 'B', 'J', and '9', using the same structure given in Figure 10.9, with N = 11.

5. Why is Q set to 1 when the quotient is zero in the hashing algorithm?

6. Modify the SYM_HASH algorithm so that it builds a hash table.

10.5
Dope Vectors

In Figure 10.9 the address field of name 'A' points to a dense list containing additional information about array A. In particular, the dense list (called a dope vector, template, or various other names) contains subscript and location information. A possible dope vector for our example is shown in Figure 10.10.

The general dope vector of Figure 10.11(a) and the example of Figure 10.11(b) show how a d-dimensional array is dope-vectored. The dope vector may be partially filled in during translation, and during execution the remainder of the parameters may be computed and stored in the dope vector.

A column-major order mapping for the dope vector of Figure 10.11 is given in terms of the parameters M_i:

$$a = \sum_{i=1}^{d} M_i (j_i - \text{LB}_i)$$

j_i = subscript of array reference

M_i = block size of multiple columns

LB_i = lower bound UB_i = upper bound

This formula is analogous to the row-major order formula given in Chapter 3.

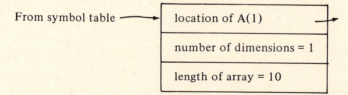

Dope vector at location 0A54

From symbol table ⟶ | location of A(1) |

| number of dimensions = 1 |

| length of array = 10 |

Figure 10.10 A dope vector for the array of Figure 10.9

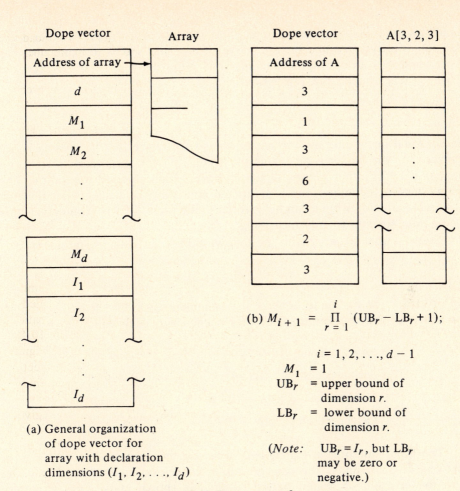

(b) $M_{i+1} = \prod_{r=1}^{i} (UB_r - LB_r + 1);$

$$i = 1, 2, \ldots, d - 1$$
$$M_1 = 1$$
$$UB_r = \text{upper bound of}$$
$$\text{dimension } r.$$
$$LB_r = \text{lower bound of}$$
$$\text{dimension } r.$$

(*Note:* $UB_r = I_r$, but LB_r may be zero or negative.)

(a) General organization of dope vector for array with declaration dimensions $(I_1, I_2, \ldots, I_d)$

Figure 10.11 Dope vector for arrays

EXERCISES

1. Compute the dope-vector parameters for an array B declared as B[10, 20, 10].

2. Compute the density of the structure in Exercise 1.

10.6
Parsing

After the compiler separates the characters in the source program into groups of characters (called *tokens*) that belong together logically, then the *parser* analyzes the tokens and groups them into syntactic structures. Another name for parser is *syntax analyzer*. The parser checks the tokens to make sure that they appear in an appropriate manner and order allowed by the source language. For example, the expression

$$X * - 2 I$$

is *not* valid in Pascal, and the parser should recognize that the syntax is not correct. As the parser is analyzing the tokens, it imposes a treelike structure that is used by other phases of the compiler, such as code generation. The many parsing techniques are best studied in a compiler design class in which a suitable amount of time can be devoted to the topic. In this section we show some data structures used in parsing.

Suppose that we are given the following PL/I assignment statement:

$$A = (A - B / A * 5) + 3$$

In order for the parser to analyze the statement correctly, it must establish an order for the operations. The order needed here is as follows:

First operation:	B / A	Divide
Second operation:	(B / A) * 5	Multiply
Third operation:	A − (B / A) * 5	Subtract
Last operation:	(A − (B / A) ∗ 5) + 3	Add

The input notation is called *infix notation*, because each operator is in between two operands. This can result in both backward and forward scanning and the creation of nested partial results, as shown above. For example, B / A is nested within the expression and yet is a partial result that must be obtained first. How are we to simplify the expression so that partial results may be obtained as we move from left to right?

We will use *postfix notation* (a form of Polish notation introduced by the logician Łukasiewicz) to overcome the problem of nesting. A postfix expression is one in which all operators follow the operands reading from left to right They are of the form (OPERANDS) . . . (OP) (OP). First the input in infix notation is converted to postfix notation. Then the postfix form can be used to generate machine-language statements corresponding to the steps needed.

We call the infix-to-postfix algorithm a *transducer*, because it converts the input into an output of different form. The transducer is guided by a precedence table, as shown in Figure 10.12.

Two stacks are operated by the transducer. Stack S_1 is a stack that stores the output of the machine, and stack S_2 is a temporary "scratchpad" memory

INPUT SYMBOL

		\<Identifier\>	=	+, −	*, /	(	)	End of statement	
	"null"	S_1	S_2	ERR	ERR	ERR	ERR	ERR	
TOP of S_2	=	S_1	ERR	S_2	S_2	S_2	ERR	U_2	
	+, −	S_1	ERR	U_1	S_2	S_2	U_c	U_2	
	*, /	S_1	ERR	U_1	U_1	S_2	U_c	U_2	
	(	S_1	ERR	S_2	S_2	S_2	U_c	U_2	

Figure 10.12 Next state function for pushdown stack transducer

used by the transducer during operation. For simplicity, only tokens that consist of identifiers, operators $(+, -, *, /)$, and parentheses will be allowed as input. The output set will consist of all of the above except parentheses.

The precedence table of Figure 10.12 may be encoded into an array that "drives" the transducer by telling it what to do with each symbol as it is input. The entries in the table have the following meaning:

S_1 = stack input onto S_1 S_2 = stack input onto S_2

ERR = error occurred, input not valid

U_1 = unstack $S_2 \rightarrow S_1$; do another comparison

U_c = unstack $S_2 \rightarrow S_1$ repeatedly until (is encountered; discard (

U_2 = unstack $S_2 \rightarrow S_1$ until S_2 empty

Let us demonstrate the operation of the transducer in Figure 10.12. The steps in Figure 10.13 show how S_1 and S_2 grow and shrink as the expression is scanned from left to right. The infix expression is shown along the bottom of each table of Figure 10.13, and two stacks, S_1 and S_2, rise above each character in the infix expression. The operation performed after reading each syntactic unit (identifier or operator) is indicated below the character read from infix form. The final result (output in postfix) is found in stack S_1 when S_2 is empty and no other inputs are made.

Let us follow through the example of Figure 10.13. The first syntactic unit is the identifier 'A'. Since the top of stack S_2 is null, the table in Figure 10.12 tells us that the operation is "stack A onto S_1." The next value is '='. Referring to the table again, we see that '=' should go onto stack S_2. Now

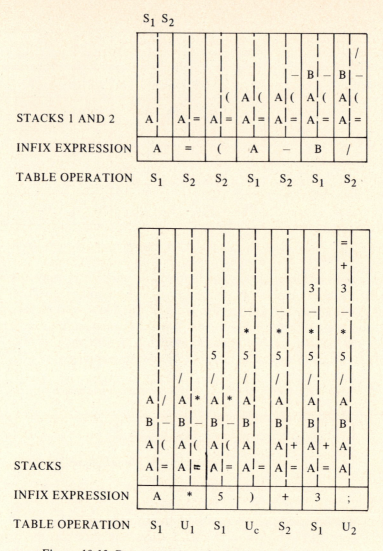

Figure 10.13 Demonstration of infix-to-postfix notation

the top of S_2 is the '=' unit, and the input is the unit '('. This results in an operation of S_2.

We continue in this fashion until S_1 contains all the input and S_2 is empty. Note that identifiers are stored on S_1 as they are read left to right. Operators, on the other hand, are always diverted to S_2 before going on S_1. The diversion to S_2 ensures that the operation will be performed in the correct order. The rules for unstacking S_2 are determined by the precedence of the operators. Bracketing is done by parentheses and causes every operator between parentheses to be pushed onto S_2 temporarily.

The result in S_1 contains no parentheses and no nesting that cannot be resolved by left-to-right scanning. Let us see how the postfix string of S_1 is interpreted so that appropriate operations can be generated in machine language:

$$S_1 : A\ A\ B\ A\ /\ 5\ *\ -\ 3\ +\ =$$

The storage element S_1 is now treated as a queue by reading from bottom to top, or as shown above, from left to right. The interpreter algorithm below uses S_2 again as a stack element for temporary storage. The interpreter must scan from left to right by storing operands in S_2 until an operator is found, then apply the operator to the top two atoms of the temporary storage stack S_2, and push the resulting value onto S_2. This is repeated iteratively until S_1 has been scanned and S_2 is empty. The interpreter algorithm is given below.

Algorithm   INTERP

1. Comments: This algorithm interprets a postfix string and performs the operations as they are given.
    Input: S_1 Input string in postfix notation
    Output: Value of expression
    Variables: S_2 Stack
                       E   An element from S_1 or S_2

2. Repeat the following until the stack S_2 is empty:
    (a)   Remove the bottom (or leftmost) element from S_1:
             E ← bottom of S_1
    (b)   If E is an identifier, then
             Push E onto S_2
          Else
             If E is an operator, then
                Perform the operator's operation on the top two elements of S_2
                Replace them with the result of the operation.

3. End INTERP

Figure 10.14 gives an example of scanning the sample string in postfix notation. Note that S_2 grows until an operator unit is encountered. The operator unit takes the top two elements off S_2 and performs an operation. The result is placed back onto S_2 and the entire procedure is repeated.

In practice, the precedence table and the syntactic units are coded to conserve storage. The codes used for identifier name and operators are called *token codes*. The tokens must be recognized as either name or operator. Recognition of tokens is sometimes done with the aid of a symbol table, as discussed previously.

We have now shown the steps in program preparation and translation. The next step is to show how data structures play a role in program execution.

				A		5					
		B	B	B/A	B/A	B/A*5		3			
		A	A	A	A	A	A	$A-T_1$	$A-T_1$	T_2	
S_2	A	A	A	A	A	A	A	A	A	A	$A=T_2$
S_1	A	A	B	A	/	5	*	–	3	+	=
Operation	Stack	Stack	Stack	Stack	Divide	Stack	Mult	Sub	Stack	Add	Assign

Figure 10.14 Interpretation of S_1 where $T_1 = B/A*5$ and $T_2 = (A - T_1) + 3$

EXERCISES

1. Write a program to recognize assignment statements as given by the parsing transducer. Assume that names are single-character alphabetic, and operators are $+$, $-$, $*$, $/$. Allow any level of parentheses nesting.

2. Write a program to interpret the output string S_1 from Exercise 1. How have you coded the tokens?

3. Devise a precedence table for the *IF* source statement of the form: IF e_1 THEN e_2 ELSE e_3. Assume that e_1 is a simple expression, such as $A < B$ or $X = Y$, so that only two names and one infix operator are allowed. The statements e_2 and e_3 can be any assignment statement.

10.7
I/O Processing

Most computers must manage the flow of data to and from peripheral devices, as well as compute arithmetic results. Unfortunately, the speed of peripheral devices differs vastly from the internal working speed of a computer. For this reason, it is necessary to provide structures called buffers for input/output (I/O). A *buffer* is a storage element used to temporarily save data that is being written or read. Once the transfer from a peripheral device has taken place, the buffer is referenced by the user's program to obtain data. Our second topic in program execution is the structure necessary for buffering.

Modern computers have the capability of simultaneously transferring data to and from main memory while processing other data. For example, a lengthy numerical calculation requires very little input/output, but very much arithmetic capability. Hence an I/O operation may take place simultaneously with calculations to increase the computer system throughput. On the other hand, I/O is typically slower than the arithmetic component of a computer, and so it is desirable to transfer large blocks of data at one time.

A method of input called *buffered I/O* will be used to demonstrate a useful data structure for increasing the processing speed of a computer system. Assume that two *buffers* (blocks of temporary storage) are used in the following manner:

> Buffer A is being used for input while
> Buffer B is being read by the program

and then, when A is full and B has been processed, their roles are reversed:

> Buffer B is being used for input while
> Buffer A is being read by the program

The ping-pong effect of these two buffers increases throughput because it allows simultaneous operation of the program and I/O (see Figure 10.15).

In Figure 10.15 we see that two areas of memory are set aside for input buffers. In addition, three pointers are needed to indicate the INPUT buffer: ACTIVE, which points to the input space, and two displacements within A and B that locate the data being accessed. The two buffers are being accessed simultaneously: A by the input processor transferring data from an outside source to location NEXT A, and B by the output processor moving the word in NEXT B to an executing program.

In Figure 10.16 the buffers have just completed a cycle. Buffer A is full and buffer B is "empty" (processed). The input is now made to B, and the processing is done on the contents of A. The ACTIVE pointer tells us which buffer is to receive the input data. Once all input has been made and all A has been processed, the cycle is completed and ACTIVE switches back to buffer A.

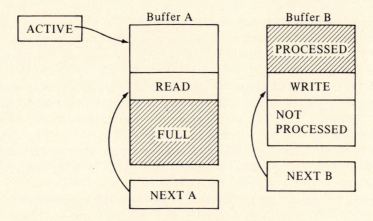

Figure 10.15 Double buffering in operation

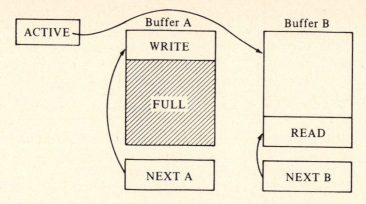

Figure 10.16 State of double buffers at beginning of next cycle

EXERCISES

1. What should be the size of buffers A and B of Figure 10.15 when I/O is performed at 20,000 bytes/second and processing is done at 40,000 bytes/second? Assume that it takes $\frac{1}{40}$ second to switch from buffer A to buffer B. (*Hint:* Let I/O time = switch time + process time for each buffer.)

2. Explain what happens if data rates exceed the processing rates in a double-buffer system.

3. Show how 3, 4, . . . and more buffers can be used in a multiple-buffer system that provides for data overrun (*data overrun* is the process of exceeding the capacity of the processor to process data as fast as it is transferred into a buffer).

10.8
Recursion

As a second example of data structures useful for program execution, we illustrate stacks used for recursion. *Recursion* is the process of nesting subprogram calls within one another. A *recursive subprogram* is a subprogram that calls itself.

The topics of recursion and I/O buffering are two features of a system that executes programs. Such features are part of the run-time environment of a computer system. The support of run-time environments is an important topic of computer science and involves extensive use of data structures. As these structures become known and understood, they are incorporated into hardware and become computer structures. Indeed, many of the following operations are operations performed by hardware instead of software.

Consider the following problem: The number 518 is stored in computer memory as a binary number (1000000110). Our problem is to separate the decimal equivalents into 5, 1, and 8 and print them out as characters. In other words, we want to convert from binary to decimal and code the decimal digits for output. A table for the ASCII-7 character set is shown in Figure 10.17.

To solve the problem by recursion, assume that division produces a remainder and a quotient. Repeated division of 518 by 10 produces first 8, then 1, then 5 as remainders.

Algorithm BIN2ASC

1. Comments: This algorithm converts a binary number into its ASCII representation, one digit at a time.

    Input: Q   Number to be converted; as number is converted, Q is a quotient
               R   Remainder, initially zero

    Output: A series of ASCII characters

2. Divide number by 10; get remainder:
    $Q \leftarrow Q / 10$; R is the remainder of the division

3. If Q is not 0, then
    CALL BIN2ASC with Q and R

4. Encode the remainder R by adding 60_8 to it:
    $R \leftarrow R + 60_8$
    Output R

5. End BIN2ASC

Initially the algorithm is called with Q = 518 and R = 0. At step 2 we divide 518 by 10 to obtain a new value for Q (Q = 51) and a remainder of 8. We then CALL BIN2ASC with these new values. Next Q is 5, R is 1, and we call BIN2ASC again. This time Q is zero and the value of R is 5, the first

Character (decimal)	ASCII-7 (octal)
0	60
1	61
2	62
3	63
4	64
5	65
6	66
7	67
8	70
9	71

Figure 10.17 ASCII-7 codebook for numerals zero through nine

digit of the number. To encode it into ASCII, we add 60_8 and obtain 65_8. An ASCII 5 is output and we return to the run-time environment with $Q = 5$ and $R = 1$. We output 61_8, an ASCII 1, and finally output 70_8, the ASCII code for 8.

The algorithm works by first dividing the value repeatedly by 10 to separate off each digit. As the algorithm exits it outputs the ASCII-encoded digits in order from left to right. Figure 10.18 traces the algorithm using the value of 518 as input.

The output is 65_8, 61_8, and 70_8, just as expected. The subtle point, however, is that the algorithm produces remainders in the exact opposite of the order desired. But with the recursion, the output order is reversed as needed.

EXERCISES

1. Write a recursive subroutine to perform the binary-to-decimal conversion of BIN2ASC.

2. How many levels of recursion will there be for an n-digit number?

3. Give a recursive algorithm and a simple example ($n = 3$), for computing $n! = n \cdot (n - 1)!$, where 0! is defined to be 1. (*Note: n!* is read n factorial.)

	$Q = 518$
	$R = 0$
Step 2	$Q = 51$
.	$R = 8$
Step 3	CALL BIN2ASC
Step 2	$Q = 5$
	$R = 1$
Step 3	CALL BIN2ASC
Step 2	$Q = 0$
	$R = 5$
Step 4	$R = 65_8$
	Output 65_8
	EXIT BIN2ASC
Step 4	$R = 61_8$
	Output 61_8
	EXIT BIN2ASC
Step 4	$R = 70_8$
	Output 70_8
	EXIT BIN2ASC

Figure 10.18 Trace of Algorithm BIN2ASC

10.9
Minimal-Path Algorithms

Computing the shortest path from one node to another in a graph has become a very important problem. Shortest-path algorithms are not necessarily difficult, but the number of computations required is often unwieldly. Applying these computational methods to problems involving transportation or communication networks would be extremely tedious without the aid of a computer.

In this section we will discuss two shortest-path algorithms. The first one, published by Edsger Dijkstra in 1959, is generally considered to be the most efficient method for finding the shortest distance between a pair of nodes. The second, published by Robert Floyd in 1962, shows how to compute the shortest path between every pair of nodes in a graph.

COMPUTING THE SHORTEST PATH
BETWEEN A PAIR OF NODES

Let us assume that the nodes in the graph are numbered 1, 2, . . . , n and that we wish to find the shortest path from node 1 to node n. The lengths of the edges in the graph are given by an incidence matrix D, in which the entries d_{ij} give the distance from node i to node j. If there is no path from i to j, let d_{ij} be ∞, meaning an infinite value or any large value that represents no path.

The algorithm involves dividing the nodes into two classes. Class 1 nodes are those whose minimum distance from node 1 has already been determined. Class 2 nodes remain to be examined. In addition to class, each node has a value. For class 1 nodes, the value is equal to the node's distance from node 1 along a shortest path. For class 2 nodes, the value represents the shortest distance found thus far. As the algorithm progresses, the value of class 2 nodes decreases if a shorter path is found. After the value of each class 2 node has been adjusted, the one with smallest value is reclassified as class 1. We will call the most recently reclassified node the *pivotal node;* its value is used to adjust the value of the remaining class 2 nodes. The algorithm terminates when node n is placed in class 1.

Algorithm   SPATH

1. Comments: Find the length of the shortest path between node 1 and node N. The graph connecting the nodes is defined in an incidence matrix D, which gives the distances between the nodes.

    Input: N   Node for whom shortest path from node 1 is to be found

    Output: VALUE[N]   Length of shortest path from node 1 to node N

Variables: D   Incidence matrix of distances between the nodes 1, 2,
                            . . . , N
             VALUE[I]   Value of node I; for a node in class 1, its value is its
                          distance from node 1 along a shortest path. For a node
                          in class 2, its value is its shortest distance from node
                          1 found thus far.

2. Place node 1 in class 1; all others in class 2.

3. Set VALUE of node 1 to zero, all others to ∞ (or a very large number).

4. Repeat the following until node N is placed in class 1:
    (a) Define the pivotal node as the one most recently placed in class 1.
    (b) Set the value of all class 2 nodes in the following way:
             If a node is not connected to the pivotal node, then its value remains the
             same
             Else
             Replace the node's value by the minimum of its current value and the
             current value of the pivotal node plus the distance from the pivotal node
             to the node in class 2.
    (c) Choose a class 2 node with minimal value and place it in class 1.

5. Output 'Shortest distance from node 1 to N is', VALUE[N].

6. End SPATH

   The algorithm stops when node N is in class 1. The value of each node
in class 1 is the length of the shortest path from node 1 to the given node.
The algorithm gives only the length of the shortest path; it does not tell what
the path is. However, if we record a node's immediate predecessor whenever
its value changes in step 4(b), we can trace the shortest path back to node
1 (see Exercise 5). The algorithm requires at most $n(n - 1)/2$ additions and
$n(n - 1)/2$ comparisons.

SHORTEST PATH BETWEEN PAIRS OF NODES

The following algorithm works with the *adjacency matrix* D that contains
elements d_{ij}, the distance from node i to node j. If there is no edge connecting
i and j, then d_{ij} is defined as ∞. The algorithm proceeds by inserting nodes
one at a time into a path only if the new path yields a shorter distance. For
example, the graph in Figure 10.19 has $d_{13} = 12$, but the distance is shorter
if the path taken from 1 to 3 includes node 2.

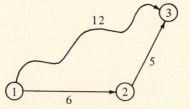

Figure 10.19 Directed graph with lengths of edges given

The algorithm produces a sequence of matrices, which we will label $D^{(0)}, D^{(1)}, \ldots, D^{(n)}$. A typical element in $D^{(k)}$ is $d_{ij}^{(k)}$, where k is some value between 0 and n. $D^{(0)} \leftarrow D$ the adjacency matrix. At any step, $D^{(k)}$ represents the distance between all pairs of nodes resulting from the possible inclusion of nodes $1, 2, \ldots, k$ in the path to make it shorter.

Algorithm   SPATH_ALL

1. Comments: This algorithm finds the shortest path between all pairs of nodes.
    Input: D   An adjacency matrix giving the distances between the nodes
                  N   Number of nodes
    Output: $D^{(K)}$ A matrix representing the distance between all pairs of nodes,
                          where nodes 1 through K are possibly included to make the
                          path shorter
                  I, J   Indexes

2. Set matrix $D^{(0)}$ to the adjacency matrix D:
    $D^{(0)} \leftarrow D$

3. Repeat the following for K = 0, 1, . . . , N − 1:
    Define $D^{(K + 1)}$ in the following way:
       Repeat for all I, J from 1 to N:
          $D^{(K + 1)}[I, J] \leftarrow$ minimum of $D^{(K)}[I, J]$
                          and $D^{(K)}[I, K + 1] + D^{(K)}[K + 1, J]$

4. End SPATH_ALL

The algorithm does not specify which paths produce the shortest distances; it merely determines the distances. To find out how the paths can be found, see Exercise 6. This algorithm requires $n(n − 1)(n − 2)$ comparisons and additions.

EXERCISES

1. Use the following nondirected graph to trace the SPATH algorithm and find the shortest distance between node 1 and node 6.

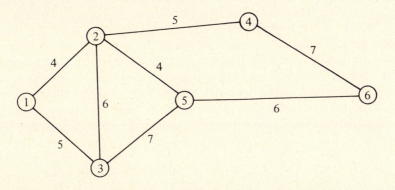

2. Use the nondirected graph in Exercise 1 and compute the shortest distance between every pair of nodes.

3. What would happen in step 4(c) of Algorithm SPATH if there were several nodes with the smallest value?

4. Write a program to implement either Algorithm SPATH or Algorithm SPATH_ALL.

5. How can Algorithm SPATH be modified so that the minimum path can be determined?

6. How can Algorithm SPATH_ALL be modified so that the minimum paths can be recorded?

10.10
B-Trees

The searching methods discussed in Chapter 8 deal primarily with data stored in internal memory. A good internal searching technique may be extremely slow when applied to data residing on a direct-access storage device. For example, a binary search requires on the average $\log_2 n$ comparisons before locating the specified element. In a disk file containing 100,000 records, this means approximately 17 disk accesses for each search, and each access involves seek time and rotational delay. A possible solution to external searching is to keep a partial index of selected records. If the records are ordered, the index determines the approximate location of the record. The indexed sequential file organization uses this strategy with two levels of indexes. The higher-level index locates the cylinder, and the lower-level index the track. Then one reads the track to find the record. For an example of this strategy, see Section 6.3.

Another organization that makes searching for and updating files relatively easy is the B-tree. The B-tree provides an index structure over the file so that insert, delete, find, and locate the next record in sequence operations are accomplished fairly quickly. The tree is constructed so that it is always balanced, no matter what the order of input data is. The original B-tree structure as described by Bayer and McCreight (see References) kept the records (or pointers to the records) in every node of the tree. In this section we will discuss the B-tree and a variation called the B^+-tree by Comer (see References). The B^+-tree is based on a suggestion by Knuth. In the B^+-tree all data records reside in the terminal nodes. The nonterminal nodes form an index over the data. Separating the records from the index facilitates the delete operation and makes it possible to link the records together for faster sequential access.

B-TREE

A *B-tree of order n* has the following properties.

1. Every node has no more than *n* immediate successors.

2. Every node, except for the root and the terminal nodes, has at least *n*/2 immediate successors.

3. The root has at least two immediate successors (but can have none).

4. All terminal nodes are on the same level and contain no keys.

5. A nonterminal node with *k* immediate successors has *k* − 1 keys.

Figure 10.20 (p. 256) shows a B-tree of order 6. The values shown in each node are the keys and each key is unique. In a node the keys are in order and there is one more branch than the number of keys. Each node can have no more than six immediate successors, so each has a maximum of five keys. Since the terminal nodes hold no data, they can be represented with nil pointers.

The general form of each node is the following:

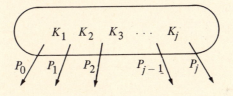

where K_i is a key value and all keys are unique.
    P_i is a pointer to an immediate successor of the node.
    j is less than or equal to $n - 1$.

If the keys are in ascending order, then $K_1 < K_2 < \ldots < K_j$. P_0 points to a node containing values less than K_1, and P_j points to a node containing values greater than K_j. For all other pointers, P_i points to a node whose values are between K_i and K_{i+1}. For example, in Figure 10.20 the pointers of the root node divide the keys into two categories: those less than 375 and those greater than 375, as determined by the two subtrees of the root.

To search a B-tree for a particular key, begin with the root node and see if the value is one of the node's keys. If it is, the search stops and is successful. If it is not, select the appropriate pointer and continue the search. If the pointer leads to a terminal node (or equivalently, if the pointer is nil), the search is unsuccessful and the value is not found. In an actual application the nodes could contain not only the keys but also the location of the record

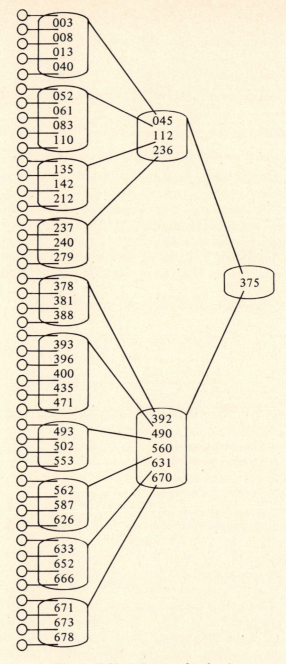

Figure 10.20 A B-tree of order 6

with the particular key. Thus when the search is successful, we would be able to retrieve the desired record.

The effectiveness of a B-tree search results from the shape of a B-tree. By definition the B-tree tends to be short (it has few levels) and wide (it has many nodes at each level). On the other hand, a binary tree tends to be long (many levels) and narrow (few nodes at each level). In a binary search, only half the remaining keys are eliminated from consideration at each step. In a B-tree search, approximately $(n - 1)/n$ of the remaining keys are eliminated at each step.

The maximum search length depends on the depth of the tree. If the terminal nodes are at level $k + 1$, the maximum search length is k. There is only one node (the root) at level 1, at least two nodes at level 2, $2\lceil\frac{n}{2}\rceil$ at level 3, $2\lceil\frac{n}{2}\rceil^2$ nodes at level 4, and so on. The notation $\lceil\frac{n}{2}\rceil$ is called the ceiling function and means "take the smallest integer greater than or equal to $\frac{n}{2}$." At level $k + 1$ there are at least $2\lceil\frac{n}{2}\rceil^{k-1}$ nodes. In a B-tree containing N keys, the number of terminal nodes is $N + 1$, so

$$N + 1 \leq 2\left\lceil\frac{n}{2}\right\rceil^{k-1} \qquad \text{or equivalently,} \qquad k \leq 1 + \log_{\lceil\frac{n}{2}\rceil}\frac{N + 1}{2}$$

This means that for a B-tree of order 256 and a file containing 100,000 records, the maximum search length is no greater than 3.

Inserting a key into a B-tree is quite simple and may cause the tree to grow upward toward the root. In Figure 10.20 there is one more terminal node than the number of keys in the entire tree. The terminal nodes are place holders where insertions may appear. For example, insertion of the key 137 causes the node to change from

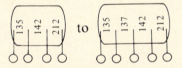

If we want to insert the value 460, however, we must use a different procedure, because the node at which the insertion should take place is already full. In this case we split the node into two halves and put the middle key into that node's predecessor node. In doing so it may be necessary to split the predecessor node and pass its key on up the tree. This process may continue until the root node has to split and a new root is defined for the tree. For example, when 460 is inserted, the node changes from

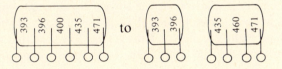

and 400 (the middle value) goes into the node's predecessor. Since the predecessor is full, it too must be split and a value is carried up to the root. After this insertion, the changed portions of the tree appear in the following way:

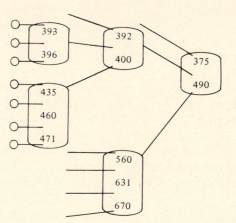

If a B-tree of order n has $i + 1$ levels (where the terminal nodes are on level $i + 1$), a key to be inserted goes into a node at level i. If that node contains n keys, split it into two parts,

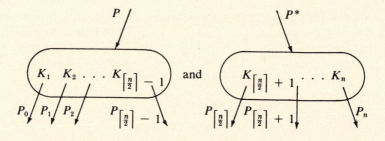

and insert key $K_{\lceil \frac{n}{2} \rceil}$ into the node's immediate predecessor. The pointer associated with the key $K_{\lceil \frac{n}{2} \rceil}$ is $P*$, the pointer to the newly created node. If the insertion of $K_{\lceil \frac{n}{2} \rceil}$ causes the predecessor node to be full, split it in a similar manner and continue the process until no more divisions are necessary. This may mean the root node splits, causing a new root to be added to the tree.

Deletion is handled in a similar manner. If a key is not at the lowest level (immediately above a terminal node), replace it with the key's successor and delete the successor. If a key is at the lowest level, removing it from a node may cause the node to contain less than $n/2 - 1$ keys. In this case we use the node's left or right brother and move the keys so that each of the two nodes are approximately the same size. In doing so it may be necessary to remove a key from the node's predecessor, which in turn may cause another removal, and so on.

The step between a B-tree and the implementation of a directory structure with the search characteristics of a B-tree is complicated by the fact that directories usually contain considerably more information than the keys and pointers. In some files, for example, a record is protected by lists of users authorized to read from or write into the record. Such extra information can be quite voluminous. And if it is stored in the nodes near the root, it can seriously affect the time to search through a node.

The nodes farther from the root also have problems. If we imagine that the records in the file are stored on a disk file, at some level a node in the index corresponds to a block of records that are transferred to and from disk as a unit. The fact that the keys are kept in order in the nodes reflects the fact that the records within each transfer block are also kept in order. Problems arise when records must be added or deleted. To facilitate the insertion of records, you can allocate some free space within each transfer block. When a record is deleted, incorporate its space into this free space. To illustrate some of the differences between a B-tree structure as defined and its application to a file structure, consider the following example.

One of the basic features of this file structure is the allocation of extra space within the nodes to keep storage management to a minimum. Inserted records may use up the empty areas, but deleted records create empty areas. If necessary, additional space can be added, but the tendency is to have the B-tree grow in width, not length. Assume that each record has a unique key that determines the ordering of the records. Retrieval of the records can be in a sequential manner or in a random manner, as determined by key value. The lowest level of the tree contains the actual records in key sequence. The higher levels of the tree contain pointers and key values to facilitate file storage, retrieval, and update. Figure 10.21 (p. 260) shows a 3-level structure. The second-level index contains the highest key value in each of its successor record blocks. It also contains a pointer to the location of each record block. Some of the record blocks may be entirely empty, and some may have space included at one end. Similarly, some of the second-level index nodes may be empty, or only partially full. The horizontal pointers on the second-level index make it possible to have sequential access to the records without having to refer to the first-level index.

Characteristics of the secondary storage device help determine the size of the nodes. At the lowest level, the size of a node is the amount that is transferred to and from main memory and depends on the length of the records and the amount of storage allocated for I/O buffers, as well as the device in which the file is stored. The lowest-level nodes are grouped so that they occupy an integral number of tracks. Normally this is a cylinder to limit the amount of head movement. To save space, keys in the index level nodes are stripped of those beginning and ending characters that do not help distinguish

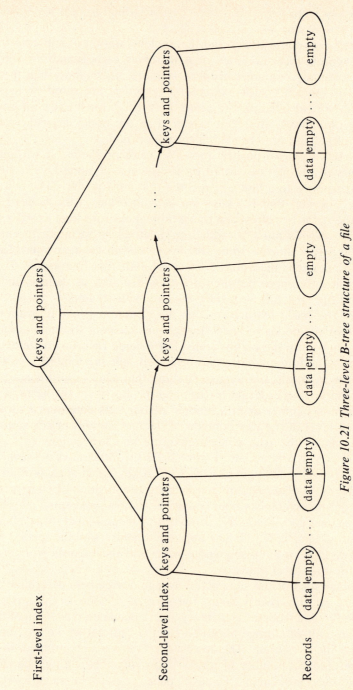

First-level index

Second-level index

Records

Figure 10.21 Three-level B-tree structure of a file

them from adjacent keys. Thus more information can be packed into a single index node. Control information is kept as part of each record node so that the records can have either fixed or variable length.

Each second-level index node points to the same number of record nodes, which may or may not be full. If an insertion causes the splitting of a record node, approximately half the data will be copied into an empty node and the insertion made. The empty node and the split node must have the same predecessor. If there is no empty node available, the predecessor node is split, and approximately half its record nodes are assigned to a new index node. This splitting will create two index nodes, each of whose record nodes are half empty, thus avoiding further splitting for new insertions.

When records are deleted, the space is reclaimed and added to accumulate free space at the lowest-level nodes. Within a node the free space is always at the end.

The above discussion pertains to the organization of a single file. In the overall system, the directory of allocated files could be kept with the same sort of B-tree structure.

B⁺-TREE

Recall that the B⁺-tree is a B-tree explicitly designed so that the terminal nodes contain the data records. The upper-level nodes contain only keys and pointers to other nodes. Figure 10.22 pictures a B⁺-tree of order 2 for a company's state sales representatives. Because of space limitations, we include only the state code, the salesperson's first name, and the number of state accounts. The actual data could contain much more information. Each nonterminal node of the tree has room for four keys (a two-character state abbreviation in this example) and five pointers.

In general, a B⁺-tree of order n has at most $2n$ keys and $2n + 1$ pointers in every nonterminal node. Every key appears at least twice in the B⁺-tree: in the original record and in one of the nonterminal index nodes of the tree.

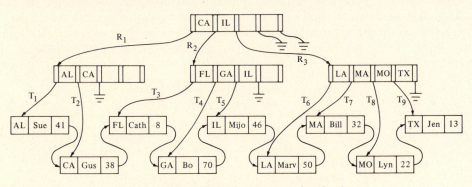

Figure 10.22 Sample B⁺-tree of order 2

The root node's leftmost pointer, labeled R_1 in Figure 10.22, points to the index node of all keys less than or equal to CA. Pointer R_2 locates the index node of all keys less than or equal to IL. The last pointer (R_3 in Figure 10.22) points to the index node of all keys greater than IL. Note that the root node has room for two more keys.

Searching the B^+-tree is similar to searching a B-tree. Looking for a particular key (say GA) in the B^+-tree of Figure 10.22, we begin at the root node. Since GA is not less than or equal to CA, but it *is* less than or equal to IL, we follow pointer R_2. This locates a B^+-tree node with pointers to the data records. In the implementation there would be a flag in each node to denote whether its pointers pointed to data or to other index nodes. To locate the GA record, we find the GA key in the B^+-tree and follow the T_4 pointer to the data record.

The organization of the B^+-tree lets us find a record quickly. At each level in the tree, we have the choice of taking one of n branches (for a B^+-tree of order n). We choose the appropriate one and eliminate the other $n - 1$ branches. For $n = 2$, this is the binary search process. For $n > 2$, the search is even faster. In general, locating a record takes at most

$$\lceil \log_n r \text{ probes}$$

where r is the number of records in the file. Table 10.1 gives the number of accesses required for a file with a certain number of records and a given B^+-tree order. Note that locating any record takes the same number of accesses, because a search must go through every level of the tree.

The order of a B-tree or a B^+-tree is usually designed around the capacity of a disk sector, track, or cylinder. This tends to make the programming rather dependent on the device. But if each node is contained on a single sector or track, it can be retrieved with only a single seek.

Figure 10.22 shows the records linked together in key sequence order. Thus, if the file is to be processed sequentially, we can follow the links to locate the next record in sequence without having to go through the index. Although the diagram shows the records in sequence, the order in which they are written on the external storage device depends on their order of insertion or input. The links from one record to the next establish the alphabetical order.

**Table 10.1   Number of Accesses to Locate
a B^+-Tree Record**

B^+-tree order	1000 records	10,000 records	100,000 records
10	3	4	5
50	2	3	3
100	2	2	3

To illustrate the effect of inserting a record into a B^+-tree, we add Hawaii (HI) to our file. We discover that it will fit in the node containing FL, GA, and IL. Since we keep the keys in order within the nodes, we must rearrange some of the keys and pointers. To add a record for Oregon (OR), we see that we must split one of the index nodes. First we insert it in the node in which it should go, arrange the data, and then divide the node into two parts. The first $n + 1$ keys become one node, the last n keys become another node, and the middle key gets replicated in the parent node. The middle key helps locate the two nodes that have split. Figure 10.23 shows the sample B^+-tree after the insertion of OR and HI. In this case, when MO moved up to the root, the root did not have to split.

Following is an algorithm for inserting a record into a B^+-tree.

Algorithm   B_INSERT

1. Comments: Insert a record into a B^+-tree. This assumes that all the keys are unique.
   Input: RECORD   The data record to be inserted
                KEY   The key field of the input record
                N   Order of the B^+-tree
   Output: Nothing

2. Find the node in the index B^+-tree that should contain the KEY and a pointer to the RECORD. Use a stack to record the path taken to reach this node.

3. If the KEY will fit into the node, then
      Place KEY in the node;
      Decide where the RECORD will be written on external storage;
      Point to the RECORD's location on external storage;
      EXIT B_INSERT

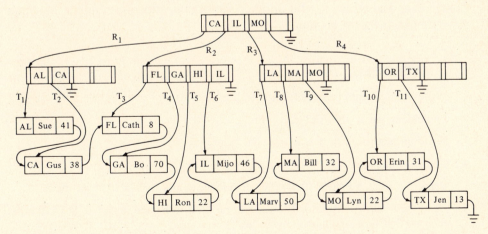

Figure 10.23 The B^+-tree of Figure 10.22 after OR and HI have been inserted

Else
  Divide the node into two parts;
  Add the KEY to one part of the node, point to its RECORD's location on external storage;
  Set aside the middle key value of the node to be moved up to its parent node

4. If there is no parent node, then
  Create a new root node;
  Insert the middle key value in the root node;
  Add pointers to the split nodes
Else
  Insert the middle key value in the parent node;
  If the parent node is already full, then
    Repeat the splitting operation growing the tree upward

5. End B_INSERT

The B^+-tree structure has become a standard for organization of files. Some dedicated database systems and general-purpose access methods are designed using the B^+-tree structure. It provides either sequential or random access to the data. As in a B-tree, the effectiveness of a B^+-tree search results from the shape of the B^+-tree. By definition, the B^+-tree tends to be short (to have few levels) and wide (to have many nodes at each level).

One disadvantage of the B-tree or B^+-tree structure is its possible waste of storage space. Every index node is a fixed size and may contain many empty entries. To save space and pack more information in a node, the keys may be stripped of those beginning and ending characters that do not help distinguish them from adjacent keys. If many deletions occur within the file, then either space may be wasted or time may be spent coalescing the index entries.

The B-tree and its variations remain an interesting structure to study. Compare and contrast it with another file organization—extendible hashing, discussed in the next section.

EXERCISES

1. Write an algorithm to perform a search of a B^+-tree.

2. Implement the B_INSERT algorithm.

3. Write an algorithm to delete a record from a B^+-tree.

4. Compare and contrast the B-tree and the B^+-tree.

5. Draw the tree in Figure 10.23 after deleting GA and then after inserting AZ.

6. Describe how the B-tree or B^+-tree can be used to handle duplicate keys, that is, records that have the same key value.

10.11
Extendible
Hashing

Extendible hashing is a method for accessing files, especially those that are constantly changing and experiencing frequent updates. It is designed to locate any data record in no more than two accesses to the external storage media containing the file. It successfully combines hashing and radix searching, and offers an alternative to the B-tree structure for files. The extendible-hashing index can expand (hence the word *extendible*) and contract as the database expands and contracts.

We have seen that static structures, such as sequential files or dense lists, require much effort or rearranging in order to add or delete elements. Adding records to a sequential file could be done at the end, but any order inherent in the data probably would not be maintained after the addition. Many insertions to an indexed sequential file may create overflow areas and lengthen search time for a scheme whose intent is to provide fast access to the data.

To motivate the extendible-hashing structure, recall that the search time for data kept in internal memory falls into three basic categories, depending on the data structure used. The time to search a sequential structure is directly proportional to n, the number of items being searched. A tree- or binary-type search has search time proportional to $\log_2 n$. A hashing scheme is best of all and ideally has search time equal to 1, regardless of the number of items being searched. Thus the fast access of hashing motivated the originators of extendible hashing to design a scheme that would allow hashing to be used to access files.

The internal-storage hashing methods that we described previously in the text begin with a hash table of a fixed size and a hashing function that depends on the size of the hash table. If the hash table's density becomes too close to 1 and there are many collisions, then there is no way to increase the size of the table without rehashing every key. As a contrast, the extendible-hashing scheme that originated with Fagin *et al.* (see References) allows the hash table to grow when necessary to provide immediate access to the data. Other similar hashing methods are expandible hashing described by Knott, dynamic hashing by Larson, and virtual hashing by Litwin (see References). In this section we discuss only extendible hashing.

The entire extendible-hashing structure is composed of pages. A *page* in the author's terminology is a fixed-size area that can contain data records or pointers to other pages. A *directory page* contains only pointers. A *leaf page* contains only keys and their records or pointers to the records.

Figure 10.24 shows the basic extendible-hashing structure. The directory may consist of one or more pages. Its entries point to the leaf pages. The hashing function used may be any one that is suitable. Fagin *et al.* recommend

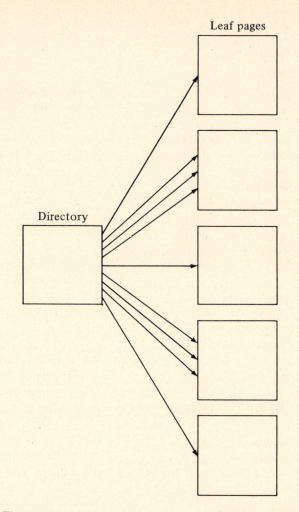

Figure 10.24 Basic data structure of extendible hashing

one selected from a universal class of hashing functions as defined by Carter and Wegman (see References). These hashing functions distribute the keys uniformly over the address space.

The basic search algorithm of extendible hashing is as follows.

1. Given a key to search for, apply the hashing function to it.

2. Use the result of the hashing function to locate an entry in the directory.

3. The entry points to a leaf page containing the key. If the key is not in the leaf page, then it is not in the file.

The two accesses to locate an item are first one access to the directory and then one to the leaf page. If the leaf pointed to the data record instead of containing the record, then there would be an additional access. The thing that makes extendible hashing work is the partitioning of the data into groups on the basis of the hashed value of their key. This partitioning is based on the same kind of distribution used in radix sorting, and is a buddy-system type of partitioning.

For example, suppose that initially we have a very small file of 16 records. Each record's key is 12 characters long, but our hashing algorithm folds the key and uses the division technique to compress the key into 32 bits. The hash value can thus be from 0 to $2^{31} - 1$, or 2,147,483,647. In a typical hashing scheme we would need a hash table with 2,147,483,647 entries—quite a large amount of space for a file beginning with only 16 records. Of course we expect the file to grow, but we could probably get by initially with a much smaller hash table. That is what extendible hashing does—it makes the hash table just as large as it needs to be and uses only a part of the hashed key's value to locate the records. Figure 10.25 shows the pseudokeys for each of the records in our sample file with 16 records. The *pseudokey* is the result of hashing the record's key. Figure 10.25 pictures only the first 5 bits of each pseudokey. Figure 10.26 shows an extendible hashing structure for the file.

In our example, each leaf page can hold 3 to 6 pseudokeys. Although it is not shown, the leaf page may also contain the records or pointers to the records associated with each of the keys. The first leaf page is pointed to by the 00 directory entry, since each of its pseudokeys begins with 00. Similarly, the second leaf page contains all the pseudokeys beginning with 01. The last leaf page is pointed to by two directory entries, 10 and 11. This means that the leaf page contains pseudokeys beginning with either 10 or 11. The small numbers above each leaf page box give the number of digits, which are the same for every pseudokey in the page. In each of the first two pages the pseudokeys thus have the same first two digits. The last leaf page's pseudokeys all begin with the same first digit—the number 1 in this example.

The way in which the extendible hashing structure is built comes from the

Record number	Pseudokey	Record number	Pseudokey
1	00101 . . .	9	00111 . . .
2	11000 . . .	10	01101 . . .
3	11110 . . .	11	01110 . . .
4	00110 . . .	12	10001 . . .
5	10100 . . .	13	01100 . . .
6	00011 . . .	14	01011 . . .
7	11001 . . .	15	10010 . . .
8	01111 . . .	16	01000 . . .

Figure 10.25 Sample file with record's pseudokeys

Leaf pages

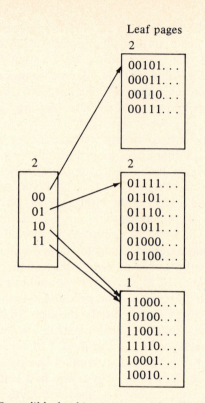

Figure 10.26 Extendible hashing structure for file of Figure 10.25

way in which records are inserted. For example, to insert a record with pseudokey 00001 . . . , we first examine the directory. The 2 outside the directory page tells us to look at the first 2 bits of the pseudokey (00 in this case). The 00 directory entry points us to the first page, where the pseudokey and its record can be inserted. If the pseudokey collides with (is equal to) another pseudokey already in the leaf page, then some predetermined collision resolution scheme must be used to locate the new record in the page.

Inserting a record with pseudokey beginning with 10 or 11 causes the third leaf page to split, since it is currently full. This creates a fourth leaf page and some rearranging of the data within the third leaf page. All records beginning with pseudokeys 10 are in one page; those beginning with 11 are in the other page. The depth of each leaf page has increased by 1. The *depth* of a page is the number of bits (or digits) that are the same for each pseudokey. Figure 10.27 illustrates the updated extendible hashing structure after the insertion of pseudokey 11100. The leaf page has been split and the directory pointers have been updated.

Inserting a record with pseudokey 01001, we see that the second leaf must split. Since its depth is already 2, the depths of the new leaves will be 3. One

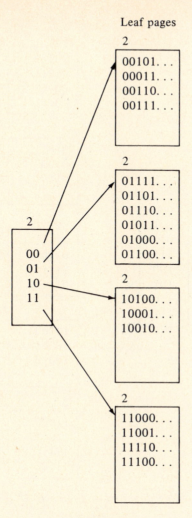

Leaf pages

2
| 00101... |
| 00011... |
| 00110... |
| 00111... |

2
| 01111... |
| 01101... |
| 01110... |
| 01011... |
| 01000... |
| 01100... |

2
| 10100... |
| 10001... |
| 10010... |

2
| 11000... |
| 11001... |
| 11110... |
| 11100... |

2
00
01
10
11

Figure 10.27 Inserting record 11100 into structure of Figure 10.26

leaf will contain pseudokeys beginning with 010; the other will have pseudokeys beginning with 011. How will the directory point to these leaf pages?
Its depth is only 2. In order for the directory to locate a leaf page of depth
3, the directory must be able to locate a pseudokey three digits long. This
means that the directory itself must double. Figure 10.28 pictures the doubling
of the directory and the insertion of pseudokey 01001. Note that now most
of the leaf pages have two directory entries pointing to them. Only the two
newly created pages have one pointer. This directory is large enough to accommodate eight leaf pages ($8 = 2^3$, where 3 is the depth of the directory).
    In an actual application the pages would be large enough to hold many

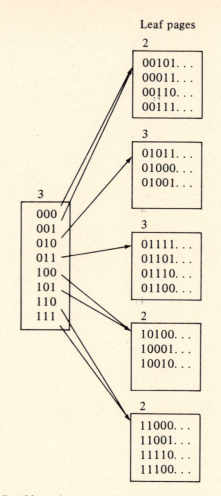

Figure 10.28 Doubling directory of Figure 10.27 by inserting 01001

more pseudokeys and pointers. We limit ourselves here for simplicity. To keep the structure balanced, we impose a loading-factor range of values for each page. When the page is almost full and exceeds the maximum loading factor, it should be split into two buddy pages. If many records are deleted from a page and its loading factor falls below a predetermined minimum, then the page could be combined with its buddy.

Searching for a particular key is similar to inserting a key. First we hash the key, use as many bits as the directory requires, and locate the leaf page containing the key. Within the leaf page we may use any search technique to locate the record. Since each page is a fixed size, a traditional hashing scheme may give us immediate access to the data record. Within the page, any collisions must be resolved by keeping the record synonyms within the same page.

Algorithms for search, insert, and delete follow.

Algorithm EH_SEARCH

1. Comments: Search an extendible hashing structured file for a record with a particular key. The algorithm assumes that all keys are unique.

   Input: KEY   Key of the record to be located

   Output: RECORD   Record containing KEY

   Variables: DIRECTORY   Extendible hashing directory for the file

       LPTR   Pointer field of DIRECTORY

       PKEY   Pseudokey value of KEY

       D   Depth of file's directory

       I   Index into directory

2. Perform the hashing function H on KEY:
   PKEY ← H(KEY)

3. Take the first D bits of PKEY and store them in I:
   I ← first D bits of PKEY

4. Get the pointer to the leaf page containing keys beginning with I:
   LPTR ← pointer field of DIRECTORY[I]

5. Search leaf page pointed to by LPTR for record with key value KEY.

6. If record not found, then
   Output 'Record not found for key', KEY

7. End EH_SEARCH

Algorithm EH_INSERT

1. Comments: Insert a record into an extendible hashing structured file. This algorithm assumes that all keys for the records are unique.

   Input: KEY   Key field of the record to be inserted

       RECORD   Record to be inserted

   Output: Nothing   Unless an error occurs

   Variables: DIRECTORY   Extendible hashing directory for the file

       PKEY   Pseudokey value of KEY

       D   Depth of file's directory

       I   Index into directory

       LPTR   Pointer to leaf page where record is to be inserted

       NEW_LPTR   Pointer to new leaf page if a split occurs

       PD   Leaf page's depth

2. Perform steps 2 through 4 of Algorithm EH_SEARCH.

3. If PKEY and RECORD will not cause the leaf page to split, then

Copy PKEY and RECORD into the leaf page pointed to by LPTR.
EXIT EH_INSERT

4. Get space for a new leaf page, pointed to by NEW_LPTR.

5. Increase the leaf page's depth by 1:
    PD ← PD + 1

6. Place the records into leaf page pointed to by LPTR or NEW_LPTR.

7. If the new leaf page's depth is greater than the depth of the directory, then
    Increase the directory's depth by 1:
    D ← D + 1
    Double the size of the directory and update pointers.
    Else
    Update the directory so that it points to leaf page LPTR and NEW_LPTR.

8. End EH_INSERT

Algorithm EH_DELETE

1. Comments: Delete a record from an extendible-hashing structured file. This al-
                gorithm assumes that all keys for the records are unique.
    Input: KEY   Key field of the record to be deleted
    Output: Nothing   Unless an error occurs
    Variables: DIRECTORY   Extendible-hashing directory for the file
                PKEY   Pseudokey value of KEY
                D   Depth of file's directory
                I   Index into directory
                LPTR   Pointer to leaf page on which record is to be inserted
                NEW_LPTR   Pointer to new leaf page if a split occurs
                PD   Leaf page's depth

2. Perform the EH_SEARCH procedure, obtaining the address of the record to be
    deleted.

3. If there is no record with key value KEY, then
    Output 'Record not found'
    EXIT EH_DELETE

4. Delete the record by making the pointer to it nil, or by making its storage area
    blank or nil.

5. If the page containing the deleted record and its buddy page can be combined into
    a page with loading factor less than the desired maximum, then
    Combine the pages and update the directory.

6. If every pointer in the directory is the same as its buddy's pointer, then
    Decrease the directory's depth by 1 and halve the directory.

7. End EH_DELETE

Simulations performed by Fagin *et al.* have shown the extendible-hashing structure to perform at least as well as or better than B-trees in terms of time to access the data. The time to access any item is the same as the time to access any other item. There is no best- or worst-case performance. A single directory access and a single leaf-page access locate the data record. If the directory is on external storage instead of in memory, then there are two I/O operations.

One disadvantage of extendible hashing is that the records are not necessarily stored in key sequence order, as the B$^+$-tree records are. Thus the B$^+$-tree offers the possibility of sequential or random access. The extendible hashing is designed for random access. To judge whether this is a true disadvantage, we have to evaluate the application using extendible hashing. If the keys in each leaf page are kept in their key sequence order, then by merging the leaf pages, we can process the records sequentially.

EXERCISES

1. Write an algorithm to make one pass over the directory, doubling its size and updating the pointers. Assume that directory entry I and its buddy I' are to point to a leaf page L that splits into L and its buddy L'.

2. Implement an extendible hashing structure for a file.

3. Perform a simulation to compare extendible hashing and B-trees.

4. Study the references given in this section and compare the other hashing techniques with extendible hashing.

A
STRING PROCESSING WITH PROGRAMMING LANGUAGES

We will show how to program the string operations performed on the expression X = A1 + B of the example in Section 2.2. This can be done in a variety of ways and with a variety of programming languages. We will illustrate only a few string processing capabilities of SNOBOL, Pascal, and PL/I. For a complete explanation of Pascal, SNOBOL, or PL/I, consult a programming text.

As you may recall, the problem is to scan the expression string and separate the variables X, A1, and B from the operators, equals sign (=), plus (+), and "end of statement." The operators and variables will be placed in a separate storage area capable of storing strings. Thus we wish to find the substrings X, =, A, 1, +, B, and the terminating blanks, and then concatenate them into the substrings X, =, A1, +, and B.

PL/I

Let us see how we would do the separation and concatenation of PL/I. There are five basic constructions in PL/I that provide all the operations necessary for concatenation, insertion, deletion, replacement, indexing, and determining of length. We give the form of each in Figure A.1.

In addition to the PL/I string functions, we will use PL/I statements for control, initialization, and assignment. You will have to get these statements from other sources. We intend here only to demonstrate string processing in a high-level language.

The program is developed top down by stepwise refinement, beginning with a gross outline and then constructing modules. Each module will perform

PL/I Function	Description	Example
A ‖ B	Concatenate strings A and B	S = 'STR' ‖ 'ING';
INDEX(S,SUB)	Integer index of first character of SUB embedded in S	N = INDEX ('ABC','BC'); result: N is 2
SUBSTR (STR,START,LENGTH)	(a) as a function SUBSTR returns the substring starting at START within STR and of length LENGTH	S = SUBSTR ('ABC',2,2); result: S is 'BC'
	(b) as a pseudo-variable SUBSTR replaces the substring within STR that starts at START and is of length LENGTH	SUBSTR ('ABC',2,2) = 'XY'; result: 'AXY'
DCL S CHAR(N) VAR;	This declares S to be a CHARacter string of VARying length (maximum N). N must be a constant.	DCL A CHAR(100) VAR; A can store up to 100 characters (or less)
LENGTH(S)	Integer valued length of S	N = LENGTH (STR);

Figure A.1 PL/I string functions

a specific function clearly defined by comments in the source code. In the final step we combine all modules into a working program. Let us start by defining the modules with the PL/I comment statement as shown in Figure A.2.

Starting with the comment modules of Figure A.2, we now begin to construct modules. The modules may contain additional submodules. We continue in this manner until we have finally constructed the refined PL/I code. The first module is shown refined in Figure A.3 (next page).

The data-declaration module shows how we would define variables for string arrays of size 10—for example, VARI(10) and OPER(10)—will hold the separated variables and operators. The CHAR attribute indicates that char-

```
SCAN: PROCEDURE OPTIONS (MAIN);
          /* DECLARE VARIABLES                           */
          /* INITIALIZE INPUT STRING AS                  */
          /*    STRING = 'X = A1 + B'                     */
          /* DELETE BLANKS                               */
          /* SEPARATE CHARACTERS                         */
          /* ADD TO VARIABLE STRING OR                   */
          /* ADD TO OPERATOR STRING                      */
          /* OUTPUT STRINGS                              */
END;
```

Figure A.2 Body of PL/I program to scan a string

acters will be stored within the arrays (strings). VAR indicates that the lengths
of strings are allowed to vary up to the maximum amount specified by the
CHAR attribute. The variables are initialized with an INITIAL parameter. For
example, ten null strings are placed in VARI, and the characters 'A', 'B',
and so on are placed in ALFA. The FLAG string is used to switch between
operator separation and variable separation.

During the course of our stepwise refinement we may need to modify the
data-declaration module to reflect changes in the program design. It is assumed
that a module flowchart has been drawn beforehand, so we have a good idea
what variables are needed.

The next module we need is the input/output code segment. Let us select
the simplest PL/I input and output statements: GET DATA and PUT DATA.
These constructions allow us to input the string by merely inputting STRING
= 'X = A1 + B';. The output will be equally simple. Figure A.4 illustrates
how this is done.

```
      /* DECLARE VARIABLES                              */
      /* INITIALIZE INPUT STRING AS                     */
      /*    STRING = 'X = A1 + B'                        */
      /* DECLARATIONS MAY BE GENERALIZED BY CHANGING    */
      /* SIZES AND INITIAL VALUES                       */
DCL VARI(10)    CHAR(2) VAR INITIAL((10)(1)' '),
                              /* VARIABLE STRING         */
    OPER(10)    CHAR(1) INITIAL((10)(1)' '),
                              /* THE OPERATOR STRING     */
    ALFA(6)     CHAR(1) INITIAL('A', 'B', 'X', '1', '+', '='),
                              /* ALPHABETICS             */
    STRING    CHAR(80) VAR,   /* INPUT TEXT              */
    WORK_STR CHAR(1),         /* TEMPORARY CHARACTER     */
    FLAG        CHAR(4) INITIAL('OPER');
                              /* VARI OR OPER FLAG       */
```

Figure A.3 The data-declaration module

(a) The Input Module

```
/*   INPUT STRING                    */
     GET DATA (STRING);
```

(b) The Output Module

```
/*   OUTPUT STRINGS                  */
     PUT DATA (VARI, OPER, STRING);
```

Figure A.4 Input/output modules

The input text may contain unwanted blanks. We will remove embedded blanks with a module that deletes blanks (Figure A.5). This module also demonstrates some very helpful string processing techniques in PL/I. Observe that string operators may be used both in a DO WHILE clause and as a string replacement operator.

The DO WHILE clause computes an index value for the first blank encountered when scanning STRING from left to right. As long as this value is not zero, the clause is executed. When all blanks have been deleted, the clause is skipped.

SUBSTR is used as a pseudovariable in the replacement step. First an index value for a blank (' ') is found, and then the blank is replaced by a null (") character. In some implementations of PL/I, this construction will not work, so we must use the statements below, which replace the statement in the DO WHILE of Figure A.5.

```
I_BLANK = INDEX(STRING,' ');
LONG = LENGTH(STRING) − I_BLANK;
STRING = SUBSTR(STRING,1,I_BLANK − 1) ‖
                  SUBSTR(STRING,I_BLANK + 1,LONG);
```

For either construction, the STRING is now packed (without blanks) and is possibly shorter. Note that we declared STRING as VARying up to 80 characters.

The heart of our PL/I example is the module that separates the character string into individual characters and places them in the appropriate VARI or

```
/*   DELETE BLANKS   */
     DO WHILE (INDEX(STRING,' ')  ¬ = 0);
         SUBSTR(STRING,INDEX(STRING,' '),1) = ";
     END;
     LONG = LENGTH(STRING);   /*   STRING MAY BE SHORTER   */
```

Figure A.5 The delete-blanks module

OPER string arrays. In Figure A.6 we see that additional submodules are needed to complete the stepwise refinement.

The VARI string array is indexed with the integer variable I_EQ, and the OPER string array is indexed with I_OP. The FLAG indicator is used to direct the program to store WORK_STR as an operator 'OPER' or as a variable 'VARI'. J is an index into the STRING string. The value of J runs the full length of STRING, which is given by LONG. In our simple example we have tested ALFA(5) and ALFA(6) for FLAG = 'OPER'. In a more difficult problem we would resort to a more sophisticated search mechanism.

For the final refinement, we specify the /* MOVE TO OPER STRING */ and /* MOVE TO VARI STRING */ submodules. We note that an operator is only one character, while a variable may be one or more characters long. Each string of one or more characters is an element of an array. The submodules must therefore be able to determine the array element index and the string index needed to concatenate characters into strings. In Figure A.7 we show how these two functions are performed.

Now that the modules have been refined, let us see how they work. The essential steps of the algorithm are: Select a character from the STRING by scanning from left to right (J = 1 TO LONG). If the character is an alphabetic symbol, concatenate it to a VARIable string until an operator is scanned. If the character is an operator, store it in the OPER string array. Each time a unit (either operator or variable) is found, increment the appropriate string array index.

The remainder of the algorithm deals with packing (deleting the blanks) and input or output. The program must have data-declaration information to tell PL/I what variables are used as string arrays and what the length of the

```
      /* SEPARATE CHARACTERS                                      */
      /* ADD TO VARIABLE STRING OR                                */
      /* ADD TO OPERATOR STRING                                   */
      /* OUTPUT STRINGS                                           */
I_EQ = 0;
I_OP = 0;
FLAG = 'OPER';
  DO J = 1 TO LONG;                     /* LONG IS LENGTH OF STRINGS  */
    WORK_STR = SUBSTR(STRING, J, 1);
    IF WORK_STR = ALFA(5) | WORK_STR = ALFA(6) THEN
      /* MOVE TO OPER STRING                                      */;
    ELSE
      /* MOVE TO VARI STRING                                      */;
  END;
END SCAN;
```

Figure A.6 The separate-characters module, including submodules

```
/*   MOVE TO OPER STRING   */
DO;
  FLAG = 'OPER';
  LOP = LOP + 1;
  OPER(LOP) = WORK_STR;
END;

/*   MOVE TO VARI STRING   */
DO;
  IF FLAG = 'OPER' THEN
    DO;
       FLAG = 'VARI';
       LEQ = LEQ + 1;
    END;
    VARI(LEQ) = VARI(LEQ) || WORK_STR;
END;
```

Figure A.7 The move submodules

strings is. You might like to compose the PL/I program and test it. The output of the variables and the operators should be the following:

$$VARI(1) = 'X' \qquad OPER(1) = '='$$
$$VARI(2) = 'A1' \qquad OPER(2) = '+'$$
$$VARI(3) = 'B'$$

SNOBOL

The SNOBOL language was developed to manipulate character strings. It contains operators to concatenate strings, perform pattern matching, and do replacement. Unlike PL/I, simple SNOBOL variables do not have to be declared. The SNOBOL language processor automatically allocates storage for strings and allows the strings to be of varying length during program execution. However, declarations are necessary if users want to set up arrays or define their own data structures (see Appendix C).

Figure A.8 (pp. 280–281) lists some operations and functions in SNOBOL. Most of these will be used in the following example, in which we input a statement like

$$X = A1 + B$$

and separate out the variables X, A1, and B, and the operators = and +. It is not our intention to teach the SNOBOL language; we wish only to illustrate some of its string-manipulation abilities. In Figure A.8 the word "pattern" denotes either a single string or several strings joined with the concatenation or alternation (|) operator.

SNOBOL Operation	*Description*	*Example*
string$_1$ string$_2$	concatenate string$_1$ and string$_2$; to denote concatenation there must be at least one blank between string$_1$ and string$_2$	'SUB' 'STRING' result: 'SUBSTRING'
subject pattern	pattern matching; result is either success (subject contains the pattern) or failure (subject does not contain the pattern)	TEXT 'BY' :S(A)F(B) result: if the string 'BY' is contained in the string named TEXT, then the next statement executed is A, otherwise the next executed is B.
SIZE(string)	gives the length of string	SIZE('BJS') result: 3
string$_1$ | string$_2$	generates a pattern which will match either string$_1$ or string$_2$	'ABC' | 'XYZ' will succeed if the string tested contains either the substring 'ABC' or 'XYZ'
var = INPUT	reads one card and stores it under the variable named "var"	TEXT = INPUT result: one card is read and can be referenced with the name TEXT
OUTPUT = string	prints one line containing the characters in string	OUTPUT = 'PAGE 1' result: writes PAGE 1
&TRIM = value	If &TRIM has a non-zero value and a card is read, trailing blanks will be deleted. Otherwise all 80 characters will be read.	&TRIM = 1 result: unless the value of &TRIM changes, all cards read will have trailing blanks deleted.

Figure A.8 Basic SNOBOL operations

SNOBOL Operation	Description	Example
subpattern = obj	replacement statement; if the subject string (sub) contains the pattern string, the first occurrence of the pattern will be replaced with the object string (obj), otherwise the subject is not changed.	'ABCABC' 'AB' = 'XY' result: 'XYCABC'
subpattern . obj	conditional value assignment; if the pattern string matches a substring of the subject, the substring matched becomes the value of obj	'X = A + B' ('+'|'−') . OP result: OP = '+'
variable = value	assignment statement	NUMBERS = '0123456789'

Figure A.8 Basic SNOBOL operations (continued)

To locate the operators and variables in our example statement, we assume that the statement has the form

$$var \ op \ var \ op \ var \ . \ . \ . \ op \ var$$

where var is a variable and op is an operator. The statement

$$X = A1 + B$$

is of the correct form.

The basic outline of our program is:

1. Input the statement.

2. Delete all blanks.

3. Remove the first variable from the statement and print it.

4. Repeat the following until the last operator is removed.
   (a) Remove an operator from the statement string and print it.
   (b) Remove a variable from the statement and print it.

To enable us to recognize variables and operators, we define these patterns: LETTERS contains all the letters of the alphabet, NUMBERS contains the digits 0 through 9, and OP contains +, −, * (multiplication), / (division), and = operators. NULL is the null character string, and BLANK is a single blank character.

```
LETTERS = 'ABCDEFGHIJKLMNOPQRSTUVWXYZ'
NUMBERS = '0123456789'
OP = '+ − */ = '
NULL =
BLANK = ' '
```

To recognize a variable as either a single letter or a letter followed by a string of letters or numbers, we define the following:

```
ALPHANUM = LETTERS NUMBERS
VARIABLE = ANY(LETTERS)(SPAN(ALPHANUM)|NULL)
OPERATOR = ANY(OP)
```

ALPHANUM is the string formed by concatenating LETTERS and NUMBERS and contains all the alphabetic characters and numeric digits. The statements above introduce the SNOBOL functions ANY and SPAN. The function ANY(string) forms a pattern that matches any single character appearing in its string argument. For example, ANY(OP) can be used to see if a string contains any of the characters + − * / or =. The function SPAN forms a pattern that will match a run of characters. For example, SPAN(ALPHANUM) will see if a string contains all letters and numbers. The statement

```
VARIABLE = ANY(LETTERS)(SPAN(ALPHANUM)|NULL)
```

defines a pattern named VARIABLE, which will succeed if the string tested contains a single alphabetic character followed by either a run of letters and/ or numbers or nothing (the null string).

We begin by setting &TRIM to 1 so that all cards read will have trailing blanks deleted. Then we input the statement, store it in TEXT, and print it.

```
&TRIM = 1
TEXT = INPUT
OUTPUT = TEXT
```

Next we remove all blanks from the line by executing the following statement. It checks TEXT to see whether it contains a blank. If so, the blank is removed (replaced with the null character) and the statement is executed again. REMOVE is the label of the statement. If a blank is found in TEXT, the match is considered successful, and the machine returns to REMOVE to execute it again. Thus, when all blanks are out of TEXT, the statement following REMOVE is executed.

REMOVE TEXT BLANK = NULL   :S(REMOVE)

Now we begin to see whether there is a variable in TEXT. If so, the string matching the VARIABLE pattern is placed in VAR. If there is no such variable found, the statement labeled ERROR is executed. ERROR is defined later.

SEARCH TEXT VARIABLE. VAR   :F(ERROR)

Next we remove the variable from the text by replacing it with the null string and then output VAR.

TEXT VAR = NULL
OUTPUT = VAR

After finding and removing a variable from the statement, we proceed to look for an operator, remove it from the string, and print it. If there is no operator, we have found the variable at the end of the string, so we stop.

TEXT OPERATOR . OPR   :F(END)
TEXT OPR = NULL
OUTPUT = OPR   :(SEARCH)

If an operator is found and printed, we proceed to the statement labeled SEARCH to look for the next variable. The following statement will print a message if an error occurs and the text does not end with a variable. The END statement halts execution.

ERROR OUTPUT = 'ERROR IN STATEMENT'
END

The complete program appears as follows.

```
LETTERS = 'ABCDEFGHIJKLMNOPQRSTUVWXYZ'
NUMBERS = '0123456789'
OP = '+ - */='
NULL =
BLANK = ' '
ALPHANUM = LETTERS NUMBERS
VARIABLE = ANY(LETTERS)(SPAN(ALPHANUM)|NULL)
OPERATOR = ANY(OP)
&TRIM = 1
```

```
         TEXT = INPUT
         OUTPUT = TEXT
      REMOVE TEXT BLANK = NULL        :S(REMOVE)
      SEARCH TEXT VARIABLE . VAR      :F(ERROR)
         TEXT VAR = NULL
         OUTPUT = VAR
         TEXT OPERATOR . OPR          :F(END)
         TEXT OPR = NULL
         OUTPUT = OPR                 :(SEARCH)
      ERROR OUTPUT = 'ERROR IN  STATEMENT'
      END
```

STRING PROCESSING IN Pascal

Some versions of Pascal extend the language for string processing by embedding strings into the language. A new type named **string** is defined in the following way. Note the use of **packed array of char.** It says that the characters will be stored contiguously in the smallest amount of space possible.

```
type
   string : packed array [1 . . 255] of char;
var
   S   :   string;                    (* Up to 255 in length        *)
   T   :   string [80];               (* Up to 80 in length         *)
```

The characters stored in strings S and T can be manipulated by string intrinsics, as given in Table A.1. Some functions are known by another name in different implementations. Table A.1 gives the names of some string intrinsics and notes some alternates. For a complete list, consult a reference manual for Pascal.

Table A.1   Pascal String Intrinsics

LENGTH(S)	(* Returns the length of S *)
POS(PATTERN, S) or	(* Returns the position of PATTERN in S *)
INDEX(PATTERN, S)	(* Returns the position of PATTERN in S *)
CONCAT(S, T) or S ‖ T	(* Returns the concatenation of S and T *)
COPY (S, LOC, LEN) or	(* Returns character substring of length LEN *)
	(* starting at position LOC inside string S *)
SUBSTR(S, LOC, LEN)	(* Returns character substring of length LEN *)
	(* starting at position LOC inside string S *)
DELETE(S, LOC, LEN)	(* Removes character substring of length LEN *)
	(* from S starting at position LOC *)
INSERT(PATTERN, S, LOC)	(* Insert PATTERN into string S at location *)
	(* LOC *)

Some examples will help you to understand these intrinsics. Suppose that we want to remove all characters from string S and make it the empty string.

DELETE(S, 1, LENGTH(S));

We can initially set S and T as follows:

S := 'ABC';   T := '123';

Then concatenate them and store the result in string M:

M := CONCAT(S, T);   or alternatively,   M := S ‖ T;

The variable M is now 'ABC123'. We might want to extract a substring from this value of M:

S := COPY(M, 3, 2);   or for some implementations,   S := SUBSTR(M, 3, 2);

and the result is S = 'C1'. We might also want to know where letter 'C' is located within S, where S is 'C1':

I := POS('C', S);   or   I := INDEX('C', S);

The result is that I is 1. Finally we can insert a substring into another string at any position. Assume that S is 'ABC'.

INSERT('5', S, 2)

makes S now 'A5BC'.

Since these Pascal string operations are so similar to the PL/I string operations already discussed, we will not repeat the solution of separating the input string into variables and operators. We leave this as an exercise. Also, if you are using a version of Pascal that does not have these string functions, we suggest that you implement them as an exercise.

APPENDIX

B

PL/I LIST PROCESSING

B.1
PL/I
Constructs

As a supplement and an illustration of how to perform list processing, we present this appendix on PL/I. You must know three basic constructions of PL/I in order to understand the list-processing capabilities of the language: how to define the fields of an atom, how to reference the fields of an atom, and how to allocate or release space for an atom. Once you know these three constructions, you can build any list, tree, or graph structure discussed in this book.

We will limit this discussion to three types of PL/I variables: pointer, numeric, and character. The pointer variables will be used exclusively for reference of an atom by its address. The numeric and character variables will be used exclusively for reference of information contained in atoms. In addition, we will use other variables to control the execution of programs.

Let us see how to define the fields of an atom. The data-aggregate declaration in PL/I provides a method of labeling fields while at the same time specifying their type—pointer, numeric, or character. An example in Figure B.1 shows how we use level numbers 1 and 2 to indicate an atom name or a field name.

It is important to understand that the declaration of Figure B.1 *does not* allocate memory space for the atom pictured in B.1(a). The declaration simply informs PL/I that *when* we allocate space for ATOM, we will refer to that space in the way shown.

Every linked list, tree, or graph that is allowed to change by updating the data structure must be anchored to a permanent HEADER atom. The PL/I compiler will provide this facility if we give it a BASED variable. This BASED variable is a pointer that emits from the HEADER (see Figure B.2).

defines a data **type** called TRICOLOR and an ordering on the scalar values named RED, WHITE, and BLUE. The implied ordering is RED < WHITE < BLUE. There is no value less than RED or greater than BLUE. The compiler assigns values to the names so that the computer can work with them.

Enumerated scalars contribute to improvements in the readability of Pascal programs. We can use meaningful names as the values for variables. Here are some examples.

For months of the year, we might form the **type**

MONTHS =
          (JAN, FEB, MAR, APR, MAY, JUN, JUL, AUG, SEP, OCT, NOV, DEC)

and for those who play card games,

    SUIT = (CLUBS, DIAMONDS, HEARTS, SPADES)

We could even give names to constants

    CONS = (ZERO, ONE, TWO, THREE)

and use them in the Pascal loop,

    for I : = ZERO **to** THREE **do**   or   **for** COLOR : = RED **to** BLUE **do**

Enumeration associates an integer constant with each named scalar. These constants cannot be changed by the program, nor can they be input or output. They are simply a device for improving the reliability of a program by improving its readability.

The second way in which programs written in Pascal are improved is through the notion of subrange. A *subrange* restricts a variable to a specified range of values. Often we can catch errors in programs by restricting the values of a variable to a subrange. The compiler generates code that checks to make sure that the subrange has not been violated. Thus

    type
      INDEX   = 1 . . 100;
      MAJOR_SUIT = HEARTS . . SPADES;

are restrictions on the types integer and SUIT. Any variable of type INDEX can assume a value only from 1 to 100. Any other value would cause an execution-time error.

D.5
Variable
Declaration

Now that we know that all variables in Pascal must be typed, how do we tell the compiler what type to associate with each variable? We say that a type has been *instantiated* when it is used in a **var** statement to associate a type with a variable. The term instantiate comes from the idea that a variable is but a single instance of a type. Thus

```
var
   A, B : integer;
```

declares that A and B are two instances of integers. Note the syntax: A colon separates the list of variables from their type, and the line is delimited with a semicolon. We can illustrate variables instantiated with the simple scalar types as follows:

```
var
   TEXT  : char;         {  TEXT is a character        }
   I     : integer;      {  I is an integer            }
   W     : 1 . . 10;     {  W is in [1, 10]            }
   FLAG  : boolean;      {  FLAG is TRUE or FALSE      }
```

Every Pascal block may have only one **var** declaration statement, so we must declare all variables in a single **var** statement. Since we can nest other blocks inside the main program blocks, we can construct collections of variables. These collections have names.

The variables declared in a **var** statement placed in the program block (the outermost block) are called *global variables* because they can be accessed by all other blocks. The variables declared in **var** statements placed inside nested blocks are called *local variables* because they can be accessed only by their local instructions.

In the following sample program, variables AVAR and BVAR are global variables, while XLOC and YLOC are local variables. We chose the suffixes VAR and LOC to imply the use of the variables as either global or local. In a program, the user may choose any appropriate name.

```
program V (INPUT, OUTPUT);
   var
      AVAR, BVAR : integer;        {  Global variables           }
procedure FIRST;
   var
      XLOC : integer;              {  Local variable             }
   begin
```

```
                                      {  Executable code goes here          }
   end;                               {  End of FIRST                       }
function SECOND: real;
   var
      YLOC  :  integer;               {  Local variable                     }
begin
                                      {  Executable code for SECOND         }
end;                                  {  End of SECOND                      }
begin
                                      {  Main program code goes here        }
end .
```

The executable code inside procedure FIRST may use (or copy or change) values stored in AVAR, BVAR, and XLOC. The instruction code inside function SECOND can access the values stored in AVAR, BVAR, and YLOC (but not XLOC). The instruction code of the main program can access the values stored in AVAR, BVAR, and the value returned by the function SECOND. The main program cannot directly access the values stored in the local variables XLOC and YLOC.

Named blocks (procedures and functions) can communicate values back and forth through a list of formal parameters. A *formal parameter* is a variable used as a place holder while the named block is being translated into machine language, and used as a value holder during execution of the named block.

For example, in the following, PARM is a formal parameter and X is the *actual value* passed from the main program block to the procedure block.

```
program CON (INPUT, OUTPUT);
   var
      X :  real;
   procedure INSIDE (PARM : real);
      var
          Y  :  integer;
   begin
                                      {  Executable code goes here          }
   end;                               {  End of INSIDE                      }
begin
   INSIDE (X);
end .
```

Note that the procedure INSIDE is executed (called) when its name is used outside the procedure and the formal parameter PARM is given a value. This happens in the main program above when it executes INSIDE and gives the formal parameter PARM a value of X. It is very important to note that the main program can access variable X only, and cannot access PARM or Y. However, procedure INSIDE can access X, Y, and PARM. But what is the difference between X and PARM?

Recall that PARM is a formal parameter, and serves only to act as a place holder. The actual value used in place of PARM is the value of X. Hence PARM cannot be changed by INSIDE, but X can be changed by INSIDE. Also note that the type of the formal parameter PARM is exactly the same as the actual parameter X.

The ability of INSIDE to change global variables such as X is called a *side effect*. Side effects are dangerous in programming because they are difficult to keep track of and can lead to errors. For this reason, and others, Pascal provides for two ways to communicate actual values to procedures.

A procedure communicates through a collection of formal parameters by either pass by value or pass by variable. In *pass by value,* the procedure is passed a copy of the value. An assignment to the formal parameter will change the copy—not the actual parameter. Thus pass by value causes no side effect on the actual parameter.

In *pass by reference* (also called pass by variable) the procedure may change the formal parameter. The altered value is then passed back to the main program. In this case, the side effect causes the actual parameter to be altered. Here is an example of both kinds of parameter passing. The **var** in the parameter list designates a pass by reference parameter.

```
program HUMMER (INPUT, OUTPUT);
   var X, Y :     boolean;
   procedure IN  (XX : boolean;        {  Pass by value                          }
                    var YY : boolean);  {  Pass by reference                      }
   begin
      YY := TRUE;                       {  Side effect on Y                       }
      X  := FALSE;                      {  Side effect on X                       }
      XX := TRUE;                       {  Change to copy of actual parameter     }
   end;                                 {  End of IN                              }
begin
   IN(X, Y);                            {  Execute procedure IN                   }
end .
```

D.7
Concatenated and Nested Types

The base types of Pascal can be used to form more powerful types by concatenation and nesting. There are two ways to do this:

1. Use built-in structured types.

2. Create your own (new) types.

We first discuss built-in structured types and then show how the programmer can create entirely new structures using the **type** statement.

STRUCTURED DATA TYPES

Pascal supports four elementary structured types.

array   { List of homogeneous elements   }
record   { List of heterogeneous elements   }
set   { Unordered collection of scalar values   }
file   { Secondary storage data structure   }

The array structure is the familiar consecutive one- or two-dimensional array. For example,

```
var
    LIST    : array [1 . . 10] of integer;
    MATRIX : array [1 . . 10, 1 . . 20] of real;
```

defines LIST as a one-dimensional array of integers, and MATRIX as a two-dimensional array of real numbers. We access these elements in a program by their name, together with an index (subscripts are in square brackets, not parentheses).

$$\text{MATRIX}[I,J] := \text{LIST}[I] * \text{LIST}[J];$$

A Pascal **record** is a collection of the same or different types. A **record** type allows nesting of simpler data structures inside other data structures. For example, R is a composite of X, Y, and Z.

```
var
    R   : record
              X  : char;              { Single character      }
              Y  : array [1 . . 5] of integer;   { Array      }
              Z  : real;
          end;
```

We call X, Y, and Z the *components* of R, and reference these components using a hierarchical dot notation.

```
R.X                    { Reference to character     }
R.Y[5]                 { Reference to fifth integer }
R.Z                    { Reference to real number   }
```

Record structures can be nested indefinitely (or as far as the compiler permits).

THE SET DATA TYPE

We can compose a finite set of base-type values using square brackets to denote a *set* in Pascal. Thus the set of the prime numbers from 1 to 10 is defined over a finite subrange of integers:

var

    PRIMES : **set of** 1 . . 10;

We reference this by filling in the set:

    PRIMES := [2, 3, 5, 7];

and can test membership in the set using the operation **in.**

    if X **in** PRIMES **then** . . .
    else . . .

Here are some additional examples of sets in Pascal.

type
  DAYS   =   (MON, TUE, WED, THU, FRI, SAT, SUN);
var
  WEEKDAY : **set of** DAYS;
  Y   : **set of** ('A' . . 'Z');   { Set of letters   }
begin
  WEEKDAY := [MON, TUE, WED, THU, FRI];
  Y   := ['A', 'E', 'I', 'O', 'U'];   { Vowels   }
end .

Sets can be unioned (added), and intersected (multiplied). Thus, if

        Y := ['A', 'B'];   then   Y := Y − ['A'];

stores ['B'] in Y. The statement

        Y := Y * ['Z'];

stores [   ] (the empty set) in Y, and

        Y := ['A'] + ['B'];

returns Y to the original set ['A', 'B'].

THE FILE TYPE

Standard Pascal stores information on secondary storage devices using sequential access. Some implementations of Pascal allow random access. Either method uses the **file** type to designate the format of a secondary storage record (not to be confused with the **record** type).

```
var
  F  :  file of record
          EMP  :  integer;
          DPT  :  char;
          Z    :  real;
          W    :  boolean;
          T    :  array [1 . . 10] of char;
        end;
```

We use a modified dot notation to access each file record component. Note the up arrow.

```
F ↑ .EMP              { Integer                              }
F ↑ .DPT              { Character                            }
F ↑ .Z                { Real                                 }
F ↑ .W                { Boolean                              }
F ↑ .T[K]             { Kth character of T                   }
```

The F ↑ designation is used to point to the record containing these values. Thus, to access information on a secondary storage file, we open the file, get the *n*th record, and then use the dot notation to access any one of the components. Pascal uses a collection of intrinsic functions to do these things. For example, consider the following program fragment for reading and printing file F.

```
var
  F  :  file of record
          EMP  :  integer;
          DPT  :  char;
          Z    :  real;
          W    :  boolean;
          T    :  array [1 . . 10] of char;
        end;

begin
  RESET(F);                       {  Open file F                        }
  while not EOF(F) do
    begin
```

```
      writeln (OUTPUT, 'Employee', EMP, 'Department', DPT);
      GET(F);                        {   Get next record of F           }
   end;
end .
```

For more information and examples of file structuring, study the chapters on files and searching.

EXTENDED DATA TYPES

Pascal makes available a second method of concatenating and nesting data types into more powerful data structures. You can use the **type** statement to define a new type according to the application's need.

```
type
   STRING = packed array [1 . . 80] of char;
```

This example shows how we could extend the structured types to include a new type called STRING. The new type can be instantiated just like any other structured type. Thus X and Y are strings of 80 characters. The word **packed** means to squeeze out any unused bits that may result from word boundaries.

```
var
   X, Y  :   STRING;               {  Two 80-character strings                   }
```

We can go even further and define subranges for use in the type statement so long as the subranges appear before they are used. The following example also illustrates the use of constants.

```
program MAIN (INPUT, OUTPUT);
const N  =  100;
type
   INDEX  = 1 . . N;
   LIST     = array[INDEX] of integer;
   BUNCH = record
              M1 : LIST;
              M2 : INDEX;
              end;
var
   A, B : BUNCH;
begin
   for A.M2 := 1 to N do
      begin
```

```
        readln (INPUT, A.M1[A.M2]);
        writeln (OUTPUT, A.M1[A.M2]);
    end;
end .
```

The **type** statement provides a means of hierarchically defining data structures by nesting. We can use these structures over and over again. Furthermore, we can build linked structures using the arrow notation to indicate indirect reference.

```
type
    NODEPTR = ↑ NODE;
    NODE = record
        INFO            : char;
        LEFT, RIGHT     : NODEPTR;  {  Pointers to NODE                     }
        end;
var
    LINK                : NODEPTR;  {  Working pointer                      }
begin
    NEW(LINK);                      {  Create a place in memory for NODE    }
    LINK ↑ .INFO        := 'A';     {  Enter letter A into NODE             }
    LINK ↑ .LEFT        := LINK;    {  Point to itself                      }
    LINK ↑ .RIGHT       := LINK
end .
```

You should supplement this brief introduction to Pascal data structures with a Pascal programmer's reference manual. However, this appendix should be enough to give you a reading knowledge of the language. Appendix A describes string manipulation in Pascal.

GLOSSARY

array. An arrangement of elements into one or more dimensions. A one-dimensional array is commonly called a vector; a two-dimensional array is called a table or a matrix.

ASCII. American National Standard Code for Information Interchange. The standard code uses 7 bits plus a parity bit to code characters for transfer between data-processing and data-communicating systems.

atom. The elementary building block of data structures. An atom corresponds to a record in a file and may contain one or more fields of data. Also called **node**.

auxiliary memory. Data storage other than main memory; for example, storage on magnetic tape or direct access devices

average. The statistical mean; the expected value

average search length. The expected number of comparisons needed to locate an item in a data structure

B-tree. A special tree structure used for random or sequential access to files

backtracking. The operation of scanning a list in reverse

backward pointer. A pointer that gives the location of an atom's predecessor

balanced sort. An external tape sort that sorts by merging together tapes, each with an equal number of strings

binary radix sort. A radix sort in which the sort radix is 2

binary search. A search method in which you begin with the middle element and discard half the list. Repeat search on sublist until you find a matching key or until dividing the list produces an empty list.

binary search tree. A binary search accomplished by storing the list in a binary tree. The tree is ordered when constructed or when insertions are made, and this facilitates the search.

binary tree.   A tree in which each node has outdegree at most 2

bit.   Either zero or one. It is derived from *bi*nary digi*t*.

block search.   To accomplish a block search, determine the block that the item might be in, then linearly search the block.

bubble sort.   Sort by exchanging pairs of keys. Begin with first pair and exchange successive pairs until the list is ordered. Also called **ripple sort.**

buffer.   A storage space used to temporarily store I/O data

buffered I/O.   A method of overlapping I/O using two or more buffers

byte.   A binary character operated on as a unit. It is usually shorter than a computer word.

cascade sort.   An external tape sort that sorts by merging strings from all but one tape onto the remaining tape. Subsequent passes merge fewer tapes until one tape contains all items.

circular list.   A linked list in which the last element points to the first one. Also called **ring.**

cluster.   See **primary cluster, secondary cluster.**

collision.   An act that occurs when two or more keys hash to the same address

column-major order.   A method of storing a two-dimensional array in which all elements in one column are placed before all elements in the next column. This method can also be used to store higher-dimensional arrays.

compaction.   Packing of data structure to make room in memory

comparative sort.   Sort by comparison of two or more keys

concatenation.   The joining together of two or more strings to form a new one

connected graph.   A graph in which it is possible to get from one node to any other node along a sequence of edges. If the graph is directed, the direction of the edges may be disregarded.

connection matrix.   See **incidence matrix.**

contiguous data structure.   See **sequential data structure.**

cycle.   A path that starts and terminates at the same node

cylinder.   The tracks of a disk storage device that can be accessed without repositioning the access mechanism

database.   A set of data consisting of at least one file. It is the information that is fundamental to a system or an enterprise.

data structure.   The relationship between data items

dense list.   A list stored in contiguous locations. Also called **linear list, sequential list.**

density.   The ratio of the amount of information to the total size of a structure

deque.   A double-ended queue. A deque allows insertions and deletions at both ends of a list.

digraph.   See **directed graph.**

directed graph.   A set of nodes and edges in which an initial and a terminal node determine the direction of the edge. An edge from node A to node B is not an edge from node B to node A.

display register.   An internal register in a CRT display terminal

distributive sort.   Sort by partitioning the list and then exchanging items until order exists between partitioned sublists

dope vector.   An atom of a linked list that describes the contents of subsequent atoms in the list

doubly linked list.   A linked list in which each atom contains two pointer fields: one points to the atom's successor and the other to the atom's predecessor. The pointer field could also be used to define two different orderings of the list.

dynamic memory management system.   A memory system that supplies variable-sized space, depending on the request

EBCDIC.   Extended binary-coded decimal interchange code. This character set uses 8 bits to represent each character.

edge.   An edge connects two nodes in a graph. An edge may or may not have direction.

empty string.   A string containing no characters (it has length zero). Also called **null string.**

external fragmentation.   At a system level, a pattern of allocated spaces interspersed with available spaces, causing a loss of usable storage

external sort.   Sort while all or part of a list is stored on an auxiliary storage device

field.   A unit of information

FIFO.   First-in, first-out queue discipline

file.   A collection of related records treated as a unit

filial set.   A collection of sons descended from a particular node in a tree

forest.   A collection of trees

forward pointer.   A pointer that tells the location of the next item in a data structure. It corresponds to a directed edge in a graph.

fragmentation.   Loss of usable memory due to checkerboarding or mismatch in fit. See **internal fragmentation, external fragmentation.**

garbage collection.   Release of unused portions of memory from a data structure to make unused areas of memory available for use

graph.   A set containing two types of objects—nodes and edges. This provides a mathematical model for data structures in which the nodes correspond to data items and the edges to pointer fields.

hashing.   A key-to-address transformation in which the keys determine the location of the data

head.   A special data item that points to the beginning of a list. A device that reads or writes data on a storage medium, such as a disk head or a read/write head on a tape drive.

heap.   An ordering on a tree such that every node is larger than its immediate successors

heap sort.   A sort that imposes a heap structure on the values as it orders them

horizontal distribution.   A method of assigning initial strings to tapes when employing the polyphase sort

Huffman tree.   A minimal-value tree. See **minimal tree, optimal merge tree.**

incidence matrix.   A two-dimensional array that describes the edges in a graph. Also called **connection matrix.**

indegree.   The number of directed edges that point to a node

index.   A symbol or numeral that locates the position of an item in an array.

infix notation.   A notation in which operators are embedded within operands

internal fragmentation.   Memory loss within a job due to mismatch between available space and requested size

internal sort.   Sort made while all items remain in main memory

IRG.   Inter-record gap

key.   One or more fields in a record that are used to locate the record or control its use

key-to-address.   See **hashing.**

leaf.   A terminal node of a tree

level.   A measure of the distance from a node to the root of a tree

LIFO.   Last-in, first-out stack discipline

linear list.   See **dense list.**

linear search.   To accomplish a linear search, begin with the first element and compare until you find a matching key or reach the end of the list.

linked list.   A list in which each atom contains a pointer to the location of the next atom

list.   An ordered collection of atoms

merge sort.   Sort which merges ordered sublists to form a larger, ordered list

minimal tree.   A tree with terminal nodes so placed that the value of the tree is optimal. See **optimal merge tree.**

multilinked list.   A list in which each atom has two or more pointers

nil pointer.   A pointer used to denote the end of a linked list

node.   See **atom.**

null string.   A string containing no characters. Also called **empty string.**

optimal merge tree.   A tree representation of the order in which strings are to be merged so that a minimum number of move operations occurs

oscillating sort.   An external tape sort that capitalizes on a tape drive's ability to read forward and backward. The sort oscillates between an internal sort and an external merge.

outdegree.   The number of directed edges leaving a node

overflow.   An act that occurs if the allotted memory for a data structure is exceeded

parsing.   The process of separating statements into syntactic units

path.   A path from node n_i to node n_j is a set of nodes $n_i, n_{i+1}, \ldots, n_{j-1}, n_j$ and edges such that there is an edge between successive pairs of nodes.

pointer.   An address or other indication of location

polyphase sort.   An external tape sort that works best with six or fewer tapes. A Fibonacci sequence of merges is established that maintains a maximum number of active tapes throughout the sort.

pop.   The act of removing an element from a stack. Also called **pull.**

postfix notation.   A notation in which operators follow the operands on which they operate

primary cluster.   A buildup of table entries around a single table location

pull.   See **pop.**

push.   The act of placing an element on a stack. Also called **put.**

put.   See **push.**

quadratic quotient search.   A hashing algorithm that uses a quadratic offset when probing subsequent table locations

queue.   A list that allows insertion of atoms at one end and deletion of atoms at the opposite end

quickersort.   Sort by partitioning a list into two sublists and a pivotal middle element. All items greater than the pivot go in one sublist and all lesser items go in the other sublist. Sublists are further subdivided until all items are ordered.

radix sort.   A distributive sort that uses a number of partitions equal to the sort radix

random access.   A method of retrieving data from a secondary storage device in which the retrieval time is independent of the location of the data. Contrast with **sequential access.**

record.   A collection of related data items. A collection of related records makes up a file.

recursion.   A reactivation of an active process; for example, a program segment that calls itself

replacement-selection.   A tournament method of sorting tape files. It produces ordered strings of various lengths that must be merged.

ring.   See **circular list.**

ripple sort.   See **bubble sort.**

root.   The node with indegree zero

row-major order.   The method of storing a two-dimensional array in which all elements in one row are placed before all elements in the next row. See **column-major order.**

scan.   An algorithmic procedure for visiting or listing each node of a data structure

scatter storage.   See **hashing.**

secondary cluster.   A buildup along a path established by a pattern in a hashing function used for table look-up

secondary memory or storage.   See **auxiliary memory.**

selection sort.   Sort by selecting the extreme value (largest or smallest) in the list. Exchange the extreme value with the last value in the list and repeat with a shorter list.

sequential access.   An access method for storing or retrieving data items that are in contiguous locations. The retrieval time of an item depends in part on how many items precede it.

sequential data structure.   A data structure in which each atom is immediately adjacent to the next atom. Also called **contiguous data structure.**

sequential list.   See **dense list.**

sequential search.   See **linear search.**

sort.   The process of placing a list in order. See **binary radix sort, bubble sort, comparative sort, distributive sort, external sort, internal sort, merge sort, quickersort, radix sort, selection sort, tree sort.**

sort effort.   The number of comparisons or moves needed to order an unordered list

spanning tree.   A subgraph of a graph with two properties: first, it is a tree, and second, it contains all the nodes of the original graph.

sparse array.   An array in which most of the entries have a value of zero

stack.   A list that restricts insertions and deletions to one end

string.   A series of characters stored in a contiguous area in memory

structure.   The organization or arrangement of the parts of an entity

subscript.   One of a set of characters used to index the location of an item in an array

synonym.   Two or more keys that produce the same table address when hashed

tail.   A special data item that locates the end of a list

terminal node.   A node of a tree that has no successors

text editor.   A program that assists in the preparation of text

threaded tree.   A tree containing additional pointers to assist in the scan of the tree

token.   A code or symbol representing a name or entity in a programming language

track.   The portion of a magnetic storage medium that passes under a positioned read/write head

traffic intensity.   The ratio of insertion rate to the deletion rate of a queue

transducer.   A device that converts information in one form into information in another form

tree.   A connected graph with no cycles. A directed tree is a directed graph that contains no cycles and no alternative paths. A directed tree has a unique node (the root) whose successor set consists of all the other nodes.

tree sort.   Sort by exchanging items treated as nodes of a tree. When an item reaches the root node, it is exchanged with the lowest leaf node.

underflow.   An act that occurs when an attempt is made to access an item in a data structure that contains no items. Contrast with **overflow.**

update.   A method by which you can modify a master file with current information, according to a specified procedure

vector.   In computer science, a data structure that permits the location of any item by the use of a single index or subscript. Contrast with a table, or matrix, which requires two subscripts to uniquely locate an item.

REFERENCES

AHO, A. B., and J. D. ULLMAN. 1979. *Principles of Computer Design*, Reading, Mass.: Addison-Wesley.

ANDERSON, M. R., and M. G. ANDERSON. 1979. "Comments on perfect hashing functions: A single probe retrieving method for static sets." *Comm ACM* 22, no. 2 (February), p. 104.

BELL, J. R. 1970. "The quadratic quotient method: A hash code eliminating secondary clustering." *Comm ACM* 13, no. 2 (February), pp. 107–109.

BUCHHOLZ, W. 1963. "File organization and addressing." *IBM syst. J.* 2 (June), pp. 86–111.

BURGE, W. H. 1958. "Sorting, trees, and measures of order." *Inform. and Control* 1, no. 3, pp. 181–197.

CARTER, J. L., and M. WEGMAN. 1977. "Universal classes of hash functions." Research Rept. RC6687, IBM T. J. Watson Research Center, Yorktown Heights, N. Y.; 1979, *Twentieth Annual Symposium on Foundations of Computer Science*, San Juan, P.R., pp. 175–182.

CICHELLI, R. J. 1980. "Minimal perfect hash functions made simple." *Comm ACM* 23, no. 1 (January), pp. 17–19.

CLAMPETT, H. A. 1964. "Randomized binary searching with tree structures." *Comm ACM* 7, no. 3 (March), pp. 163–165.

DAY, A. C. 1970. "Full table quadratic searching for scatter storage." *Comm ACM* 13, no. 8 (August), pp. 481–482.

DAYKIN, D. E. 1960. "Representation of natural numbers as sums of generalized Fibonacci numbers." *Journal London Math. Soc.*, no. 35, pp. 143–160.

DODD, G. 1969. "Elements of data management systems." *Computing Surveys* 1, no. 2 (June), pp. 117–133.

DWYER, B. 1981. "One more time—how to update a master file." *Comm ACM* 24, no. 1 (January), pp. 3–8.

FAGIN, R., J. NIEVERGELT, N. PIPPENGER, and H. R. STRONG. 1979. "Extendible hashing—a fast access method for dynamic files." *ACM Transactions on Database Systems* 4, no. 3 (September), pp. 315–344.

FLORES, I. 1960. "Computer time for address calculation sorting." *J. ACM* 7, pp. 389–409.

FLORES, I. 1967. "Direct calculation of k-generalized Fibonacci numbers." *Fibonacci Quarterly* 5, no. 3, pp. 259–266.

GOTLIEB, C. C. 1963. "Sorting on computers." *Comm ACM* 6, no. 5 (May), pp. 194–201.

GULL, W. E., and M. A. JENKINS. 1979. "Recursive data structures in APL." *Comm ACM* 22, no. 2 (February), pp. 79–95.

HARRISON, M. C. 1971. "Implementation of the substring test by hashing." *Comm ACM* 14, no. 12 (December). pp. 777–779.

HIBBARD. T. N. 1962. "Some combinatorial properties of certain trees with applications to searching and sorting." *J. ACM* 9, no. 1 (January), pp. 13–28.

HIRSCHBERG, D. S. 1973. "A class of dynamic memory allocation algorithms." *Comm ACM* 16, no. 10 (October), pp. 615–618.

HOARE, C. A. R. 1961. "Algorithms 63 'Partition' and 64 'Quicksort.'" *Comm ACM* 4, no. 7 (July), p. 321.

HOGGATT, V. E., JR. 1968. "A new angle on Pascal's triangle." *Fibonacci Quarterly* 6, no. 4, pp. 221–234.

HOOKER, W. W. 1969. "On the expected lengths of sequences generated in sorting by replacement selecting." *Comm ACM* 12, no. 7 (July), pp. 411–413.

INGLIS, J. 1981. "Updating a master file—yet one more time." *Comm ACM* 24, no. 5 (May), p. 299.

JOHNSON, L. R. 1961. "An indirect chaining method for addressing on secondary keys." *Comm ACM* 5 (May), pp. 218–222.

KNOTT, G. D. 1971. "Expandable open addressing hash table storage and retrieval." *Proc ACM SIGFIDET Workshop on Data Description, Access, and Control*, pp. 186–206.

KNUTH, D. E. 1973. *The Art of Computer Programming*, 2d ed. Vol. 1, pp. 78–96. 435–455. Reading, Mass.: Addison-Wesley.

KNUTH, D. E. 1973. *The Art of Computer Programming*, 2d ed. Vol. 3, Chapters 5 and 6. Reading, Mass.: Addison-Wesley.

LARSON, P. "Dynamic hashing." *BIT* 18 (1978), pp. 184–201.

LEE, J. A. N. 1974. *The Anatomy of a Compiler*, 2d ed., New York: Van Nostrand Reinhold.

LEWIS, T. G. 1981. *Pascal Programming for The Apple*, Reston, Va.: Reston Publishing Company.

LEWIS, T. G. 1981. "Simulation of perfect hashing functions." Report, Department of Computer Science, Oregon State University, Corvallis, Oregon.

LEWIS, T. G., and B. J. SMITH. 1979 *Computer Principles of Modeling and Simulation*. Boston, Mass.: Houghton Mifflin.

LITWIN, W.   "Virtual hashing: A dynamically changing hashing." *Proc. Very Large Data Bases Conf.* Berlin, 1978, pp. 517–523.

LORIN, H. 1975.   *Sorting and Sort Systems.* Reading, Mass.: Addison-Wesley.

LORIN, H. 1971.   "A guided bibliography to sorting." *IBM Systems Journal* 10, no. 3, pp. 244–254.

LUM, V. Y., P. S. T. YUEN, and M. DODD. 1971.   "Key-to-address transform techniques: A fundamental performance study on large existing formatted files." *Comm ACM* 14, no. 4 (April), pp. 228–239.

MARTIN, W. A. 1971.   "Sorting." *Computing surveys* 3, no. 4 (December), pp. 147–174.

MARTIN, W. A., and D. N. NESS. 1972.   "Optimizing binary trees grown with a sorting algorithm." *Comm ACM* 15, no. 2 (February), pp. 88–93.

MAURER, W. D. 1968.   "An improved hash code for scatter storage." *Comm ACM* 11, no. 1 (January), pp. 35–38.

MAURER, W. D., and T. G. LEWIS, 1975. "Hash Table methods." Computing Surveys 7, No. 1 (March), pages 5–19.

MILES, E. P. 1967.   "Generalized Fibonacci numbers and associated matrices." *Amer. Math Monthly*, no. 67, pp. 745–757.

MINKER, J., *et al.* 1969.   "Analysis of data processing systems." Technical Report, pp. 69–99. College Park: University of Maryland.

MORRIS, R. 1968.   "Scatter storage techniques." *Comm ACM* 11, no. 1 (January), pp. 38–44.

OMEJC, E. 1972.   "A different approach to the sieve of Eratosthenes." *The Arithmetic Teacher* 19, no. 3 (March), pp. 192–196.

PRICE, C. E. 1971.   "Table lookup techniques." *Computing Surveys*, 3 no. 2 (June), pp. 49–65.

RADKE, C. E. 1970.   "The use of quadratic residue research." *Comm ACM* 13, no. 2 (February), pp. 103–105.

RAMAMOORTHY, C. V., and Y. H. CHIN. 1971.   "An efficient organization of large frequency-dependent files for binary searching." *IEEE Trans. Comp.* C-20, no. 10 (October), pp. 1178–1187.

SCHAY, G., and W. G., SPRUTH. 1962.   "Analysis of a file addressing method." *Comm ACM* 8 (August), pp. 459–462.

SCOWEN, R. S. 1965.   "Algorithm 271, Quickersort." *Comm ACM* 8, no. 11 (November), pp. 669–670.

SHELL, D. L. 1959.   "A high-speed sorting procedure." *Comm ACM* 2, no. 7 (July), pp. 30–32.

SHELL, D. L. 1971.   "Optimizing the polyphase sort." *Comm ACM* 14, no. 11 (November), pp. 713–719.

SHEN, K. K., and J. L. PETERSON. 1974.   "A weighted buddy method for dynamic storage allocation." *Communications of the ACM* 17 10 (October), pp. 558–562. Corrigendum, 1975. *CACM 18*, 4 (April), pp. 202.

SHORE, J. E. 1975.   "On the external storage fragmentation produced by first-fit and best-fit allocation strategies." *Communications for the ACM 18*, 8 (August), pp. 433–440.

Smith, B. T. 1970.   "Error bounds for zeros of a polynomial based upon Gerschgorin's Theorems." *J. ACM* 17, no. 4 (October), pp. 661–674.

Sprugnoli, R. 1977.   "Perfect hashing functions: A single probe retrieving method for static sets." *CACM 18*, no. 11 (November), pp. 841–850.

Sussenguth, E. H., Jr. 1963.   "Use of tree structures for processing files." *Comm ACM* 6, no. 5 (May), pp. 272–279.

Van Emden, M. H. 1970.   "Increasing the efficiency of Quicksort, Algorithm 402." *Comm ACM* 13, no. 11 (November), pp. 693–694.

Windley, P. F. 1960.   "Trees, forests and rearranging." *British Comput. J.* 3, no. 2, pp. 84–88.

Wirth, N. 1976.   *Algorithms + Data Structures = Programs*. Englewood Cliffs, N.J.: Prentice-Hall.

ANSWERS TO SELECTED EXERCISES

Note: In this section exercises are identified by chapter and section. Thus 2.1.3 is Chapter 2, Section 1, Exercise 3. Many of the questions have a variety of answers as a result of the assumptions used and the particular implementation chosen. For example, there are several possible answers to a question that has a program or an algorithm for an answer. All of these possible answers may be correct. Therefore your answer may not exactly match the answer given here.

CHAPTER 1

1.1.2. Your discussion should include examples of increasing the length of a list by appending to the ends and by insertion in between entries. Alternatively, a list can shrink by deleting from the ends or in between the entries. When a list is ordered, it is our responsibility to maintain the order when we insert a new entry.

1.1.6. Pascal has the following types of data structure: array, record, set, and file.

CHAPTER 2

2.1.1. Locating the first record requires a pointer to its beginning. Using that pointer, plus the location of the last character of the first record, we can determine the length of the first string. The second string begins immediately after the end of the first.

2.2.2. One possibility (there are many others) is to use a storage method in which a dollar sign ($) separates strings, and we include a pointer to the first string. To concatenate the two strings:

1. Locate an area not being used.

2. Locate the first string and copy it into the area, but do not copy the $ character.

3. Locate the second string and copy it into the area following the first string; do copy the $ character.

The original strings would remain intact.

2.2.5. The algorithm must scan the string one character at a time and separate the identifiers X, A1, and B from the operators = and +. The scan builds a substring by concatenating characters until an operator (+ or =) is encountered. Blanks are removed from the substrings delimited by the operators, thus isolating the identifiers X, A1, and B.

2.2.8. Decide the characters that could be used to terminate the word 'AN' and then write the algorithm to test for them. For example, a space, a period, or a comma could mark the end of a word. Are there any other terminating characters? If 'AN' is not at the beginning of the string, it must be preceded by a blank or possibly another delimiting character, such as a right parenthesis.

2.3.1. Each record is 40 characters long.

2.3.4. Any algorithm that moves a string in memory risks copying over another string or itself accidentally. Your algorithm must protect the storage areas being manipulated. One protective algorithm works as follows: Create a temporary string long enough to hold the longest string being moved; call this TEMP. Suppose that we are to move string SOURCE to string area DESTINATION. Since DESTINATION may actually contain parts of SOURCE, we must copy SOURCE by moving it to TEMP:

$$\text{TEMP} \leftarrow \text{SOURCE}$$
$$\text{DESTINATION} \leftarrow \text{TEMP}$$

This destroys anything previously in DESTINATION, but guarantees that SOURCE is protected. It is possible to go directly from SOURCE to DESTINATION even if the areas overlap. You must be careful about which end of the string you move first.

CHAPTER 3

3.1.1. 5 bits, since $2^5 = 32$. In general, $\lceil \log_2 n \rceil$ bits are needed to address n locations. $\log_2 8192 = \log_2 2^{13} = 13$; $\lceil \log_2 10{,}000 \rceil = 14$.

3.1.3. $C[i,j]$ is stored in location

$$b + 4(j - 1) + i - 1$$

where b is the location of $C[1,1]$.

3.1.5. The pi-product computes the elements to be skipped to arrive at the proper plane. The summation generalizes plane-skipping to the n-dimensional case. The j_n term indexes into the plane.

3.1.6. $\text{Location} = \sum_{r=1}^{n-1} (j_r - l_r) \prod_{s=r}^{n-1} (u_{s+1} - l_{s+1} + 1) + j_n - l_n + 1$

3.1.7. We assumed uniform probabilities, or that $p_k = 1/n$ for every entry. We also assumed that n is a constant, but as you can see, n changes every time an insert or deletion takes place. If n were to fluctuate uniformly randomly from zero to a maximum, say max, we would compute the average over all possible lengths by summing all values of n.

$$m_i = \frac{1}{\max} \sum_{n=0}^{\max-1} \frac{n+1}{2} = \frac{\max+1}{4}$$

$$m_d = \frac{1}{\max+1} \sum_{n=0}^{\max} \frac{n-1}{2} = \frac{\max-2}{4}$$

3.1.8. 229

3.1.15. $(n-1)/2$, where n is the number of entries in the list.

3.2.2. The header is an identifier for the list. In systems in which the elements of the list may move around dynamically as a result of update activity, the header finds the actual location of the data. Without a header, the contents of the list would have to occupy predetermined locations in memory. Where memory is dynamically allocated, if the list becomes empty, only the header is left. Thus it is still possible to use the list for further insertion and deletion.

3.2.5. Your algorithm may search the entire book one name at a time, in which case you will have to search half the directory (on the average). Your algorithm might take advantage of the alphabetical order by first examining the names at the top and bottom of each page. If the sought-after name does not fall between the top and bottom names, you need not search the page further. A further refinement might be to select any page in the book. If the name you want comes before the names on the randomly selected page, you need not look in the later portion of the book. Repeat the random selection, discarding a part of the book each time. Finally, you will isolate the desired page. This method is quite rapid, surprisingly, and is analogous to browsing. The simplest algorithm is as follows, where TARGET is the phone number being looked for.

1. Let P point to the first directory entry.

2. Repeat the following until P is nil or the phone number in the entry pointed to by P equals TARGET:
    Set P to point to the next entry in the directory.

3. If P is nil, then
    Output 'Name not found'
    Else
    Output entry pointed to by P.

3.2.6. Examine m atoms, where

$$m = \sum_{k=1}^{n} k \cdot p_k \quad \text{and} \quad p_k = \frac{1}{n} \quad \text{for } k = 1, 2, \ldots, n$$

For large values of n, $(n + 1)/2 \approx n/2$.

3.3.3. A ring structure has no nil pointer indicating the end of the list. We must therefore save the address of the first atom so that it may be used as a check in place of the nil pointer of a simple linked list. In deletion, we must be careful not to delete the HEADER.

3.3.4. The total number of bits in each atom of the data structure is $240 + 30 + 16 * 2 = 302$. The number of information bits is $240 + 30 = 270$. Counting the two header pointers, the density of a list of n accounts is $270 n / (302 n + 32)$. For 100 accounts, the density is 0.89; for 500 accounts the density is 0.89. For large values of n the density is approximately $270 n / 302 n = 0.89$.

3.3.5. A dense list is easily backtracked by decrementing its index variable. A linked list with forward pointers cannot be backtracked without modification. One way to do so is to include a sign bit in the pointer field such that a $+$ indicates that the pointer is pointing forward and a $-$ indicates that the pointer is pointing backward. While scanning forward, we reverse the FORWARD pointer by pointing it to the atom preceding the atom in which it is kept, and set its sign to $-$. In this way we turn the singly linked list inside out while scanning. If we wish to backtrack, we follow the pointers that have a $-$ sign, since they point to the way back. As we backtrack, we reverse the pointer and set the signs to $+$. In a doubly linked list, we merely follow the BACKWARD pointers.

CHAPTER 4

4.1.1. Regardless of the size allocated for the queue, there is always a chance that the queue will overflow.

4.1.3. Job PR73 underflows.
Job XA41 is inserted at TAIL $= 1$.
Job BR11 is inserted at TAIL $= 2$.
Job XA41 is deleted leaving HEAD $= 2$, TAIL $= 2$.
Job BR11 is deleted leaving HEAD $=$ nil and TAIL $=$ nil.

4.1.5. The values of $z = 5$, $\rho = 4/7$ are plugged into the formulas.

$$l = \frac{\lambda}{\mu - \lambda} = \frac{4}{7 - 4} = 1.33 \qquad k = \frac{\log 1/20}{\log 4/7} = 5.35$$

Thus if the queue length is 6 or more, overflow should occur less than 5% of the time.

4.1.7. Overflow never occurs in a linked list unless the computer system is unable to provide free space. This is one of the advantages of a linked list. We handle underflow as we did with a dense list. The central problem with a linked list implementation of a queue is that of accessing the HEAD and TAIL. We must either search the full length of the list to find the TAIL (when a single link is used) or employ a doubly linked list structure that contains BACKWARD pointers. If we implement the doubly linked list structure as a ring, the HEADER atom always points to the HEAD of the list (FORWARD) and to the TAIL of the list (BACKWARD).

4.2.1. The modification is made by changing the sizes of the increment and decrement steps. TOP ← TOP + 10 and TOP ← TOP − 10.

> PUSH: 1.   If TOP = 0 (stack empty), then TOP ← 1.
>
> 2.   TOP ← TOP + 10
>
> 3.   IF TOP > N, then write "overflow."
>
> 4.   Else insert new item in locations TOP, TOP + 1, . . . , TOP + 9.

> POP: Only statement 4 changes:

$$TOP \leftarrow TOP - 10$$

4.2.3. The operands X, A, B, C, Y, and P are placed on the operand stack in the order in which they are input (left-to-right scan). The − operator is performed first, followed by the ÷ operator and the two × operators. The addition operation is done last.

> Start: X − A
> B ÷ C
> (B ÷ C) × Y
> (B ÷ C) × Y × P
> Done: (X − A) + [(B ÷ C) × Y × P]

4.2.5. In an extreme case, the first job stacked will never be output. It will be doomed to the stack base. A queue, on the other hand, guarantees that a job will be processed after an average wait of $1/(\mu - \lambda)$ time units. A queue prints output in the same order in which the jobs finish.

4.2.7. The answer to this question depends on the application. For an application requiring fast access to a stack of a known maximum length, the dense list is best. If the size of the stack is variable and may fluctuate greatly, however, the linked list is better. Which storage structure is easier to implement?

4.3.1. 1. Do only one of the following:

>   (a) If END1 is nil and END2 = 1, then overflow, else
>   END1 ← END2 − 1.
>
>   (b) If END1 = 1, then overflow, otherwise END1 ← END1 − 1.

>   2. Insert atom at END1.

>   This algorithm will not work for insertion at END2, although that
>   algorithm is smaller.

4.3.3. Approximately $(n − k)/2$.

4.3.5. A doubly linked deque is identical to the structure in Figure 3.18,
in which the header atom points to one end of the deque and the tail
to the other end.

CHAPTER 5

5.1.2. Paths from B to E containing no cycles: Cycles starting at B:

Paths from B to E	Cycles starting at B
B–E	B–A–C–B
B–D–E	B–C–D–B
B–C–D–E	B–D–E–B
B–A–C–D–E	B–A–C–D–E–B
	B–A–C–D–B
	B–C–D–E–B

The graph is connected.

5.1.4. Level 1: Y
Level 2: X, W, M, N
Level 3: A, Z
Level 4: B

5.1.6.
Node	Indegree	Outdegree
M	0	2
N	1	0
O	1	3
P	1	0
Q	1	1
R	1	0
S	1	0

5.1.8. {A, B, C, D, E, F}

5.2.3. Terminal nodes have nil pointers for both L and R. To find the root:
It is either at index 1 by convention, or pointed to from the anchor
or header. If the structure is a forest, then the anchor or header must
point to the root of each tree in the forest. True, the roots all have
indegree zero, but there may be other (unused) indexes in the structure. These would also have indegree 0, and so, perhaps, invalid
node names should be put in the structure for those indexes.

5.2.5. The repeated numbers could be inserted in the tree, of course. This would increase the depth of the tree and decrease the speed of update and search. Alternatively, we could add a repetition field to the fields in each atom. Let REP be a number indicating the number of times each number occurs in the list.

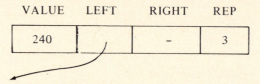

When deleting a number, we decrement the REP count and actually release the node only if REP = 0. When inserting, we either create a new node with REP = 1 or, if it exists, we increment REP.

5.2.10. The minimum number of levels is 5; the maximum is 25. In general, the minimum number of levels is $\lceil \log_2 (n + 1)$ and the maximum is n.

5.3.1. NLR order: A, B, C, G, D, E, F, H
LNR order: C, G, B, A, E, D, H, F
LRN order: G, C, B, E, H, F, D, A

5.3.3. NLR: M, X1, P1, P2, X2, X3, P3, P4, P5
LNR: P1, P2, X1, X2, P3, P4, P5, X3, M
The NLR order is related to a top-down listing of the original. LNR order is related to a bottom-up listing of the original. The original tree could be scanned from left to right by a left (L), middle (M), and right (R) scan algorithm called the NLMR order as follows: M, X1, P1, P2, X2, X3, P3, P4, P5. This results in the same order as NLR order. LMNR order and LNMR order produce different sequences of nodes, however.

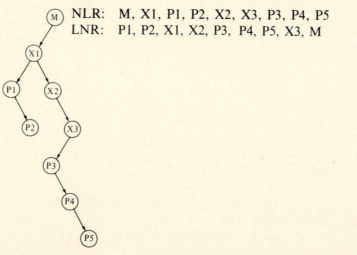

NLR: M, X1, P1, P2, X2, X3, P3, P4, P5
LNR: P1, P2, X1, X2, P3, P4, P5, X3, M

5.3.5. See the solution to Exercise 5.3.3.

5.3.7. Algorithm THREAD

  1. Comments: Scan a binary tree threaded in NLR order.
      Input: P   Pointer to tree node; initially P points to the root node
      Variables: LEFT   Pointer to node's left successor
                 RIGHT   Pointer to node's right successor

  2. Repeat the following until P is nil:
      Output node pointed to by P.
      If node's left branch is a thread,
         Set P ← RIGHT
      Else
         Set P ← LEFT

  3. End THREAD

5.3.9. The tree is easily backtracked because threads provide BACKWARD pointers where needed. The threads destroy the definition of a tree, however, because they provide alternative paths to some nodes. Therefore the threads make possible cycles in the graph. Threads change the range of permissible values for L and R pointers, or call for flags to distinguish threads from valid pointers or the nil pointer.

5.3.14.

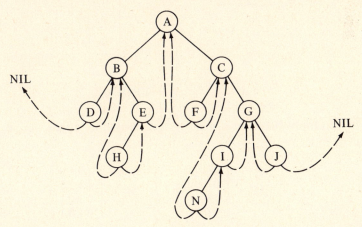

5.4.1. Visiting the nodes, as required in this exercise, you will discover a Hamiltonian path. Originally, this puzzle was in the form of a 12-sided solid. Each corner was named after a city. The Hamiltonian path was not so easily discovered by looking at the three-dimensional solid.

5.4.3. TABULAR METHOD

A:	B	C	D	E	F
B:	A	-	-	-	-
C:	D	-	-	-	-
D:	B	-	-	-	-
E:	B	-	-	-	-
F:	C	-	-	-	-

LIST-POINTER METHOD

index

A	1
B	6
C	7
D	8
E	9
F	10
–	11

INITIAL

1	B
2	C
3	D
4	E
5	F
6	A
7	D
8	B
9	B
10	C

TERMINAL

5.4.4. In a graph there are $n(n - 1)/2$ possible edges; in a digraph, $n(n - 1)$.

5.4.5.

$$M^2 = \begin{bmatrix} 0 & 1 & 0 & 0 \\ 0 & 1 & 0 & 1 \\ 1 & 0 & 1 & 0 \\ 1 & 0 & 1 & 0 \end{bmatrix} \qquad M^3 = \begin{bmatrix} 1 & 0 & 1 & 0 \\ 1 & 1 & 1 & 0 \\ 0 & 1 & 0 & 1 \\ 0 & 1 & 0 & 1 \end{bmatrix}$$

5.4.7. The scan order is F, C, D, B, A, E.

5.4.9. A path of length j shows up in M^j as a nonzero entry. If the path begins at node A and returns to node A in j steps, it will appear as a nonzero entry on the diagonal of M^j. To find cycles in a graph with n nodes, we must check the diagonals of M^j for $j = 2, 3, \ldots, n$.

5.4.11. The graph can be threaded so that each node is reached one time only by following the threads. Alternatively, the nodes of a graph can be deleted from the graph as we visit them to guarantee loopfree scanning. This has to be done carefully, however, so that we do not break a chain of pointers leading to other nodes.

5.4.14. A graph corresponding to Figure 5.28 is shown below. The problem is to cross each bridge only once. To solve the problem, define an IN counter and an OUT counter for each node (A, B, C, and D). In tracing a path of the graph, increment the IN counter each time you enter a node, and increment the OUT counter each time you leave a node. To cross all the bridges only once, you must have OUT − IN = 0 for each node. In solving this problem, consider whether or not there is a solution. If there is, give it. If there is no solution, show why not.

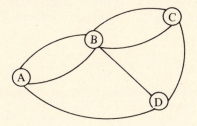

CHAPTER 6

6.1.2. The tape drive requires time to accelerate from an idle state to a full-speed running state. The time to accelerate is represented by the nonrecorded gap between blocks.

6.1.3. At 1600 bits/inch and 10 records/block, 32% of the tape will be blank. At 1600 bits/inch and 20 records/block, 19% will be blank. At 6250 bits/inch and 10 records/block, 65% of the tape will be blank. At 6250 bits/inch and 20 records/block, 48% will be blank.

6.2.2. (a) The 1980 census
       (b) An airline reservations file
       (c) Card catalog of a library (author, title, subject)
       (d) Your checking account file
       (e) Social Security files

6.3.1. See Glossary and Index.

6.3.3. Normally we hash on only one key. There are no hash functions that relate an alternate key to the hashed value of the primary key. Multiple keys are usually handled by building indexes for alternate key fields.

   Another alternative is to use all the key fields as input to the hash function. This is no more complicated than most hashing techniques, but to retrieve a record one would require knowledge of all the key fields. Usually we know one key (for example, the name) and want to retrieve all records with that key—without knowing values for alternate keys (for example, the address).

6.3.5. The algorithm must maintain the order shown. This causes problems when one inserts a record because the file may expand. Since this is essentially a block-search mechanism, consult Chapter 8. Typically, one maintains an overflow area for records that cannot be inserted because of the ordering and lack of space in the prime file area. Periodic maintenance is carried out to expand the file and clear out the overflow area.

6.3.7. A collection of overflow buckets may be set aside and used exclusively for overflow. Alternatively, the overflow may go into the next contiguous bucket. In either case, the file-search mechanism takes more time to retrieve overflowed data. If the record is not in the file, it takes a long time to verify that fact.

CHAPTER 7

7.1.1. We illustrate a selection sort that sorts the list in ascending order. The first five passes are:

⑴				
19	⑪			
26	26	⑲		
43	43	43	㉑	
92	92	92	92	㉖
87	87	87	87	92
21	21	26	43	87
38	38	38	38	43
55	55	55	55	55
11	19	21	26	38

7.1.2. The minimal sort effort for a list of 1000 items is 999. You must look at every item at least once, and if they are in the proper order already, you can do this in 999 comparisons. For unordered lists, the minimal average effort is $1000 \log_2 1000$. Some require more comparisons, some fewer.

7.2.3. Let $X[1] \ldots X[N]$ be the original list to be sorted. The MERGE algorithm will be called as a subroutine repeatedly to merge ordered substrings of X. We begin by calling MERGE with $X[1]$ for A and $X[2]$ for B. The resulting string C can be copied into $X[1]$ and $X[2]$, thereby giving an ordered pair. The second call of MERGE is with $X[3]$ for A and $X[4]$ for B, and so forth. We index through the list X until we have ordered all the pairs. The next call of MERGE uses

X[1] and X[2] for A and X[3] and X[4] for B. The merged quadruple C replaces X[1], X[2], X[3], X[4]. We index through X until all ordered pairs have been merged into ordered quadruples. We continue in this manner, doubling the length of the input substrings A and B, until the entire list is ordered.

7.2.4. The discussion on sorting does not consider the size of each record and the corresponding cost of moving long records. Names or alphabetic strings tend to be much longer than their sort keys. In considering a sorting method, we should stress the importance of average number of moves and length of records. Very long records should be separated from their keys by constructing a directory that contains keys and pointers to the master records. Thus we move only the key and associated pointer, not the entire record.

7.2.6. HEAPSORT requires as many comparisons as there are edges in the tree structure representing a given pass of the algorithm. Thus in this case there are $n(n - 1)/2 = 15$ comparisons. The number of exchanges is smaller, however, amounting to nine exchanges. The QSORT algorithm, however, bogs down in splitting the list repeatedly because it cannot find elements to exchange. The number of comparisons is roughly $n(n - 1)/2 = 15$ also, but the overhead in splitting requires much more effort.

7.3.2. String 1: 10, 20   length = 2
String 2: 15, 25, 30   length = 3
String 3: 12   length = 1
String 4: 2   length = 1

You can compute the minimal binary tree by grouping the string lengths as follows.

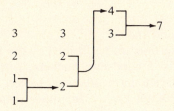

The tree has a value of $2(3) + 2(2) + 3(1) = 13$, where level numbers are put in parentheses. The value of 13 represents the number of comparisons performed in merge-sorting the numbers.

7.3.9. (a) $t = 4$, $k = 8$. Then

$$a_k^{(i)} = a_{k-1}^{(i)} + a_{k-2}^{(i)} + a_{k-3}^{(i)}$$

Thus $a_8^{(0)} = a_7^{(0)} + a_6^{(0)} + a_5^{(0)} = 105 + 57 + 31 = 193$

$a_8^{(1)} = 24 + 13 + 7 = 44$

$a_8^{(2)} = 37 + 20 + 11 = 68$

$a_8^{(3)} = 44 + 24 + 13 = 81$

CHAPTER 8

8.1.2. A FORTRAN function for computing L, given c_i and p_i is as follows:

```
        FUNCTION AVEL (N,C,P)
        DIMENSION C(N), P(N)
        SUM = 0.0
        DO 1 I = 1,N
            SUM = SUM + C(I) * P(I)
    1   CONTINUE
        AVEL = SUM
        RETURN
        END
```

8.2.3. Uniformly random keys are used to access a file whose keys are, themselves, uniformly distributed.

8.2.4. When we assume that there is no free space within the list (it is dense), the average insert causes half the list to move. The binary search locates items quickly, but does not make updating this list any easier.

8.2.5. The access frequency ordering must be extremely skewed to perform like a binary search table. The ordered table is unequaled in its rapid response to searches that reveal *no* key match. When no match is made, the access frequency table will search the entire table.

8.2.8.
$$L_{\text{block}} = L_b + L_w = \log_2(b + 1) - 1 + (s + 1)/2$$

Since $b = n/s$,

$$L_{\text{block}} = \log_2\left(\frac{n + s}{s}\right) + \tfrac{1}{2}(s - 1)$$

8.3.1. See Glossary and Index.

8.3.3. Search and insertion are performed by the linear probe hashing function. The file is divided into buckets, with overflow going into the next bucket. Deleted records are marked 'DELETED' so that the hashing chain is not broken for other records. The file is rewritten to remove deleted records only when the ratio of deleted records to total file capacity reaches a threshhold of $\sqrt{n}/n$, where n = file size.

8.3.5. The average number of probes is

$$2 = \frac{1}{1 - \dfrac{100{,}000}{n}} \qquad \text{or} \qquad n = 200{,}000$$

CHAPTER 9

9.1.2. In either system we may maintain an AVAIL list and an ALLO-
       CATED list if we reserve a few words in each block of memory for
       pointers and LENGTH attributes. An AVAIL-HEADER and AL-
       LOCATED-HEADER provide the starting point for two linked lists.
       A request is filled by scanning the AVAIL list (which links together
       all free space). A memory release is accomplished by inserting the
       freed block into the AVAIL list. If a block is not in the AVAIL list,
       it is in the ALLOCATED list. To provide a variety of block sizes,
       coalesce adjacent blocks when possible. In the fixed-block scheme
       of Figure 9.2, it doesn't make sense to coalesce blocks. There is
       always a finite, fixed number of blocks of fixed sizes. For the variable-
       block-size scheme of Figure 9.1, one should specify the allocation
       algorithm more explicitly. For example, use First Fit or Best Fit.

9.1.6. Available space: 20   15   10   5
       Requests: 15   10   5   15
       Best Fit: Allocates 15, leaving 20, 10, 5 available
                          Allocates 10, leaving 20, 5 available
                          Allocates 5, leaving 20
                          Allocates 15, leaving 5
       First Fit: Allocates 15 of 20, leaving 5, 15, 10, 5
                          Allocates 10 of 15, leaving 5, 5, 10, 5
                          Allocates 5, leaving 5, 10, 5
                          Cannot fill request for 15

9.2.1. The header information depends on the structure of the available-
       space list and the space-management algorithms used. Some possi-
       bilities for header information are the following: a pointer to the first
       available space, the total space available for allocation, the smallest
       block of space available, the largest block of space available, and
       statistical information about occupancy. There might be a separate
       list for each size of eligible block sizes with a header for each one.
       This assumes that there is only a small number of distinct sizes.

9.3.1. Yes. The Buddy system may not always be able to keep free blocks
       in buddy positions. For example, in Figure 9.3 the contiguous spaces
       at locations 8 and 16 are adjacent, but are not buddies. Thus they
       cannot be coalesced into a larger block that could satisfy a request
       for 16 words.

       We could measure external fragmentation as the ratio of allocated
       space to total space. If there are no requests for the unallocated
       space, then this measure is inappropriate. We could measure internal
       fragmentation as the ratio of requested and allocated space to total
       allocated space.

9.3.2. A multilevel system (shown here) may be more useful in the algorithm
       than the tree structure in Figure 9.3.

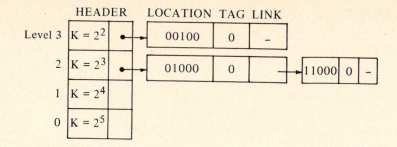

9.4.2. The 2 system is generated from

$$F_n = F_{n-1} + F_{n-3}$$

$F_0 = 0$	$F_5 = 3$
$F_1 = 1$	$F_6 = 4$
$F_2 = 1$	$F_7 = 6$
$F_3 = 1$	$F_8 = 9$
$F_4 = 2$	$F_9 = 13$

9.4.3. See Section 9.5.

9.5.1. Given n and k, we start with initial values for $F_0 = 0, F_1 = 1, \ldots, F_{k+1} = 1$ and apply the formula repeatedly to get the remaining $(n - k + 1)$ numbers. For $k = 1$, we can do this iteratively or recursively. In PL/I we could write this iteratively as:

```
F1 = 1;                           /* for K = 1 and any value of N */
F2 = 1;
DO I = 1 to N;
   F3 = F2 + F1;
   PUT LIST (F3);
   F1 = F2;
   F2 = F3;
END;
```

Recursively in PL/I for $k = 1$:

```
FN: PROCEDURE (N)  RETURNS (FIXED BINARY) RECURSIVE;
DCL N FIXED BINARY;
IF N < 3 THEN
   RETURN (1);
ELSE
   RETURN (FN(N - 1) + FN(N - 2));
END FN;
```

9.5.3. This example is almost completed in the text. The block sizes are $F_j = (m/n)j$ for $j = 0, 1, 2, \ldots, n$, and satisfy

$$F_{j+1} = 2F_j - F_{j-1}$$

This is similar to the Buddy system (Section 9.3.). If there are n block sizes in the system, keep n AVAIL lists—one for each block size. When you are splitting a block, the difference equation tells you into which AVAIL list the pieces go. Similarly, the difference equation tells you which AVAIL lists must be searched for adjacent blocks to be merged. In this system, a block splits into three pieces. It takes three adjacent blocks of the right size to coalesce into a larger block.

9.5.4. The optimal block sizes are determined by solving the state equation for F_{j+1}. Since $cdf(z)$ is

$$cdf(z) = \int_0^z \frac{e^{-x/m}}{\sigma m} \, dx = (1 - e^{-z/m})/\sigma$$

Therefore $F_{j+1} = F_j + m \left[\exp\left(\frac{F_j + F_{i-1}}{m}\right) - 1\right]$

This nonlinear equation cannot easily be solved by analytical means. Numerical solutions give F_j when $F_0 = 0$, $F_n = m$.

CHAPTER 10

10.1.1. Assume that the header is named HEAD and contains FIRST_ATOM_PTR, a pointer to the first atom in the matrix. Assume also that each matrix atom contains a VALUE, a ROW_NUM, a NEXT_ROW_PTR, a COL_NUM, and a NEXT_COL_PTR. The algorithm to search the matrix follows.

1. Get the FIRST_ATOM_PTR from HEAD and locate the first VALUE. Check its ROW_NUM. If ROW_NUM $< i$, then follow NEXT_ROW_PTR until ROW_NUM $\geq i$ or NEXT_ROW_PTR = HEAD.

2. If ROW_NUM $> i$ or NEXT_ROW_PTR = HEAD, then A[i, j] is not in the data structure, so A[i, j] = 0.

3. If ROW_NUM $= i$ and if COL_NUM $< j$, then follow NEXT_COL_PTR until COL_NUM $\geq j$ or NEXT_COL_PTR = HEAD.

4. If COL_NUM $> j$ or NEXT_COL_PTR = HEAD, then A[i, j] is not in the data structure, so A[i, j] = 0.

5. If COL_NUM $= j$, then A[i, j] = VALUE.

10.1.3. A sparse vector may be stored by recording the index of nonzero entries along with the value in a linked list. For example, A[1], A[9], A[17] . . .

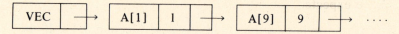

A three-dimensional array requires three links and three index values in each atom. A four-dimensional array requires four links and four index values in each atom. An m-dimensional array requires m links. Therefore the density of an $n \times n \times n \times \ldots n$ array in m space is

$$d = \frac{n^m}{(2m + 1)n^m}$$

Where we have assumed that each pointer field takes up half a word,

$$d = \frac{1}{2m + 1}$$

The break-even point is $d \cdot n^m$ nonzero entries. For a three-dimensional matrix, each atom requires seven words and d is 1/7. If the matrix's density is $> 1/7$, then the dense list saves storage. The break-even point for higher-dimensional matrices is lower.

A hashing function could be devised as follows:

$$H(i,j) = (i - 1)n + j$$

where n is the dimension bounds. This hash function results in the same waste as the original data. Suppose we know that the array will never exceed αn^2 nonzero entries. We need a hash table with load factor of α. A possible hash function now is

$$H(i,j) = \alpha((i - 1)n + j)$$

Collisions are handled by some offset mechanism, and a table of αn^2 locations is allocated instead of n^2 locations.

10.2.2. The INTENSITY field can be expanded to 2 bits. The code is

$$00 = \text{off} \quad 01 = \text{green} \quad 10 = \text{blue} \quad 11 = \text{red}$$

10.2.4. A polar coordinate CRT locates a point by a radius r and an angle a. Let a point on the complex plane be re^{ia}, where $i = \sqrt{-1}$. You can locate the real and imaginary components as in Exercise 10.2.3.

You can convert Cartesian coordinates into polar coordinates and vice versa by simple transformations. Thus it is possible to display Cartesian coordinates on a polar coordinate CRT.

10.3.1. The expected internal fragmentation is $\overline{w}$.

$$\overline{w} = \int_0^{40} \frac{(40 - x)e^{-x/5}}{5}dx + \int_{40}^{80} \frac{(80 - x)e^{-x/5}}{5}dx$$

You can calculate the expected external fragmentation by estimating how many strings will exceed 80 characters.

10.3.3. The method is naive in that it does not cope with lines of 80 characters or more. It causes fragmentation, as shown in Exercise 10.3.1. It is not easy to do inserts and maintain order in the table. Also this method does not address the problems of managing the original data file, which is probably stored on an external medium such as tape or disk.

10.4.2. The study should use $-\alpha^{-1}\ln(1 - \alpha)$ as the benchmark when computing the average comparisons to insert a symbol.| The formula $1/(1 - \alpha)$ is appropriate when look-up alone is performed.

10.4.3. The binary search tree is built in LNR order from the input which is 'A', '5', '3'. Since '5' > 'A' and '3' < '5', the tree is:

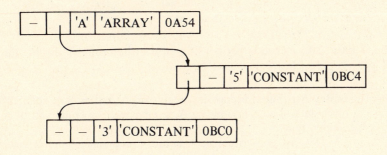

10.4.4.   fold   mod 11
        'B' = C2 → C + 2 = E   →   3
        'J' = D1 → D + 1 = E   →   3
        '9' = F9 → F + 9 = 8   →   8
        The folding operation is simple addition mod 16.

10.4.5. Q is used to resolve collisions when they occur. If Q were zero, then the algorithm would be stuck on the first collision, unable to try another location in the hash table.

10.5.2. The data uses $10 \cdot 20 \cdot 10 = 2000$ words. The dope vector uses $2 + 3 + 3$ words for address, d, M_1 to M_3, I_1 to I_3. If we agree that $M_1 = 1$ always, we need not store it. Therefore seven words are needed in the dope vector. The density is $2000/2007 \approx 0.99$.

10.6.1. See the example at the end of Section 2.2. Also see Appendix A.

10.6.3. The precedence table must be expanded to recognize IF, THEN, and ELSE. Since a true or false condition is tested, we must insert a branch on false in place of the THEN. Unfortunately we do not know the destination of the branch on false until we reach the ELSE clause. A marker must be placed on the stack so that we can come back and insert the destination later.

10.7.1. The process time for a buffer load is equal to the size of the buffer, for example, z divided by the processing speed: $z/40,000$ second. The switch time is given as $\frac{1}{40}$ second. Therefore the I/O time is

$$T_{I/O} = \frac{1}{40} + \frac{z}{40,000}$$

The time to fill a buffer is given by the size of the buffer divided by the I/O rate: $z/20,000$ second. We now solve for z.

$$\frac{z}{20,000} = \frac{1}{40} + \frac{z}{40,000}$$

$$\frac{z}{40,000} = \frac{1}{40}$$

$$z = \frac{40,000}{40} = 1000 \text{ bytes}$$

10.7.2. If data rate exceeds processing rate either one of three things happens, given buffers A and B.

1. Buffer A is filled; processing begins on A; buffer B is filled. If there is more data, it goes into buffer A, overwriting the initial part of A. It may happen that
   a. Data is overwritten in A before processing of the original data is complete.
   b. Processing of the original data in A is finished just ahead of the overrun situation about to occur. Most likely, overrun will occur in buffer B.

2. If there is more data to go into buffer A, the system may stop transfer until processing in A is completed.

3. Instead of switching between 2 buffers, we can use n buffers organized in a ring. This system will prevent overrun only if the n buffers have more aggregate capacity than a double buffer system. Having more than 2 buffers can simplify the programming checks for impending overrun.

10.7.3. Additional buffers can be used to overcome the inevitable overflow. See Chapter 3 for formulas for computing overflow. Instead of switch-

ing to a second buffer, we switch to buffer 3, then 4, . . . until we
catch up with the data flow.

10.8.1. A Pascal solution follows:

```
procedure BIN2ASC (Q: integer);
var
   X: integer;
   NEWQ: integer;
begin
   if Q > 0 then
      begin
         NEWQ := Q DIV 10;
         BIN2ASC (NEWQ);
         writeln (48 + (Q MOD 10));         { Add octal 60 (48 decimal) }
      end
end;
```

10.8.2. For an *n*-digit number, BIN2ASC is invoked *n* + 1 times. There are
n recursive calls plus the initial procedure call.
10.9.1. 14
10.9.3. It takes longer to find the shortest distance to node *n*.
10.9.5. Each time a node's value changes, record the pivotal node used to
calculate the node's value. At the end trace backward from node *n*
to node 1.
10.10.6. One record from a set of records with the same key can be placed
in the B-tree. The other records with the same key can be linked
together and pointed to by the record in the B-tree.

INDEX

access
 direct, 133
 random file, 131
 sequential, 126
 sequential file, 131
access frequency, 178
access time, main memory, 128
address, 1, 2
 disk home, 127
 disk track, 127
ADD_SPARSE, 227
adjacency matrix, 252
ALGOL, 10, 49
algorithm, 12
 ADD_SPARSE, 227
 ASCII code conversion, 249
 B_INSERT, 263
 BINTREE, 291
 BIN2ASC, 249
 BUILD_TREE, 97
 circular queue update, 69
 delete dense list, 39
 DELETE_CIRC_Q, 69
 DELETE_FIRST, 107
 DELETE_LINK, 53
 DELETEQ, 64
 DELETE_QUEUE, 65
 EH_DELETE, 272
 EH_INSERT, 271

EH_SEARCH, 271
expression evaluation, 77
extendible hashing update, 271–272
FIND, 54
GRAPH_SCAN, 118
HEAPSORT, 151
INSERT_CIRC_Q, 69
insert dense list, 39
INSERTQ, 64
INSERT_QUEUE, 65
INSERT_THREAD, 106
linear quotient hashing, 199
LINK_INSERT, 52
LNR_DELETE, 188
LNR_SCAN, 102
LQHASH, 199
MERGE, 147
minimal path, 251–252
minimal spanning tree, 115
MINSPAN, 115
POP, 79
PUSH, 79
QSORT, 157
queue update, 64
quickersort, 157
RADIX_SORT, 145
replacement selection, 162
SCAN, 276
SCAN_THREAD, 105

algorithm (*continued*)
    search dense list, 37
    shortest path, 251–252
    SPATH, 251–252
    SPATH_ALL, 253
    stack update, 79
    SYM_HASH, 237
    threaded tree, 105, 106
    tree scan, 102
    tree sort, 151
    update dense list, 38
ALLOCATE, 288–289
allocated/free list, 31, 34
alternate path, 85
ancestor, 88
APL, 49, 207
    memory management, 212
arithmetic expression, 101
    scanning, 74
array, 1, 21, 34, 240, 296
    column-major order, 42
    location of element, 43
    one-dimensional, 42
    Pascal, 311
    row-major order, 42
    SNOBOL, 296
    sparse, 225–228
    three-dimensional, 42
    two-dimensional, 4, 42
←, 38
ASCII, 15, 249
atom, 34, 61, 123
attribute list, 236
auxiliary storage, 122
available-space list, 213
average number moves during insertion, 46
average queue length, 70
average search length, 178, 179, 200
average string length, 162
AVL tree, 190

backward pointer, 58
balanced merge sort, 167
balanced tree, 190
BASED variables, 288–289
BASIC, 24
best fit, 209

binary digit, 15, 123
binary radix sort, 145
binary search, 181–183
binary search tree, 186, 291–293
binary sort tree, 96
binary tree, 90
    dense list, 94
    linked list, 95
    LNR-order, 99
    LRN-order, 99
    NLR-order, 99
    NRL-order, 99
    RLN-order, 99
    RNL-order, 99
    scanning, 102
B_INSERT algorithm, 263
BINTREE algorithm, 291, 294
BIN2ASC algorithm, 249
bit 15, 123
    escape, 229
bits per inch (bpi), 125
block
    fixed-size, 208
    variable-size, 208
block search, 183–186
boolean, 306
bpi, 125
branch, tree, 88
Bridges of Königsberg, 120
B-tree, 134, 254–264
    search, 257
B$^+$-tree, 261–264
bubble sort, 142, 146
bucket, 140
Buddy system, 214–217
buffer, I/O, 162, 246
buffered I/O, 247
buffering, double, 247
BUILD_TREE, 97

cascade merge sort, 167
c-way merge, 164
Central Processing Unit, 62, 122
char, 306
character sets
    ASCII, 15
    EBCDIC, 15
choice of data structure, 11

Cichelli, R. J., 203
circular list, 58, 63, 68
classification of data structure, 12
clustering
    primary, 196
    secondary, 197
COBOL, 24, 30, 42, 49, 202, 206
collisions, 140, 195
column-major order, 42
compaction, 209
comparative sort, 142
compiler, 29, 100
compound statement, 304
computer
    CPU, 62, 122
    execution-time stack, 73
    hardware for, 122
concatenation, 28, 310
connected digraph, 85
connected graph, 85, 114
connection matrix, 113
contiguous storage, 34
copy (substring), 29
COPY, 298
CPU, 62, 122
CRT terminal, 30, 228
cycle, 85, 114
cylinder, 127, 134

data base, 14, 130
DATA function, 299
data set, 130
data structure, 1, 3, 14, 306
    choice of, 11
    classification of, 12
    Pascal user-defined, 314–315
    scan, 178
    SNOBOL user-defined, 299-301
    user-defined, 299, 314
declaration, 308
delay
    rotational, 128
    seek, 128
delete, 9, 36, 61
    dense list, 39
    deque, 81
    linked-list, 53
    queue, 65

stack, 74
string, 28
tree, 107
DELETE_CIRC_Q, 69
DELETE_FIRST, 107
DELETE_LINK algorithm, 53
DELETEQ, 64
DELETE_QUEUE, 65
deletion rate, queue, 70
dense list, 34, 61
    delete algorithm, 39
    insert algorithm, 39
    search algorithm, 37
    storage of tree, 94–95
    update algorithm, 38
density, 45, 141, 226
    tape, 125
depth of a page, 268
deque, 61, 81–83
    delete, 81
    input-restricted, 81
    insert, 81
    output-restricted, 81
    overflow, 81
descendant, 88
device, peripheral, 122
dictionary, 134
digit-analysis hashing, 192
digraph, 85, 111, 113, 118
    connected, 85
    weighted, 110
direct access file, 133
directed graph, 85
directed tree, 86, 91
directory, 122–123
    telephone, 35
directory page, 265
directory tree, 136
disconnected graph, 85
disk, 5, 122, 126
    cylinder, 127
    file, 131
    head, 127
    home address, 127
    I/O, 128
    overflow, 128
    pack, 126
    read/write head, 127

rotational delay, 128
   search, 128
   seek, 128
   seek delay, 128
   sort, 174
   track, 127
   track address, 127
display graphics, 228–232
display register, 229
distributive sort, 142
division hashing, 140, 193
dope vector, 234, 235, 239, 240–241
double buffering, 247
doubly linked list, 58, 81
drive, tape, 123
drum, 122, 126, 128
dynamic memory-management system,
  210
   levels, 210
dynamic structure, 11

EBCDIC, 15
edge, 84
editing text, 30–33, 232–235
effort
   search, 180
   sort, 143
EH_DELETE algorithm, 272
EH_INSERT algorithm, 271
EH_SEARCH algorithm, 271
empty string, 21
end-of-file mark, 124
end-of-reel mark, 125
enumeration, 306
environment, execution, 24
escape bit, 229
Euler, Leonhard, 120
evaluation of arithmetic expressions,
  74, 77, 101
even parity, 123
execution time, 24
execution-time stack, 73
expression, arithmetic scanning, 74,
  101
EXPR_EVAL, 77
extendible hashing, 134, 265–273
external fragmentation, 208
external sort, 142, 160–177

Fibonacci memory-management system, 217–219
Fibonacci sequence, 170, 220–223. See
  also polyphase sort
field, 34, 130
FIFO, 61
file, 5, 34, 129–141, 233
   direct access, 133
   disk, 131
   indexed sequential, 135
   multiple key, 131
   random access, 131
   sequential access, 131
   single key, 131
   system, 136
   tape, 131
   tree organization, 136
   update, 132, 134
   user, 136
filial set, 88, 164
FIND algorithm, 54
first fit, 209
first in, first out, see FIFO
fixed length, 31
   records, 21, 22
fixed-size block, 208
folding, 140, 193
forest, 88
formal parameter, 309
FORTRAN, 10, 12, 24, 42, 49, 206
forward pointer, 58
fragmentation
   external, 208
   internal, 208
free-space list, 31
free-space management, 81
frequency, access, 178
function, 303
   minimal perfect hashing, 204

game tree, 6
gap, 128
   inter-block, 124
   inter-record, 124
garbage collection, 209, 235
global variable, 308
graph, 7, 110–121
   connected, 85, 114

graph (*continued*)
    cycle, 85, 114
    directed (digraph), 85
    disconnected, 85
    edge, 84
    node, 84, 111
    nondirected 84, 111
    number of paths, 113
    path, 85, 111
    path length, 113
    representation of, 112
    scan, 117
    shortest path, 111, 252
    storage of, 111
    weighted, 110, 114
graphics, display, 228–232
GRAPH_SCAN, 118

Hamiltonian path, 111
hardware, 122
hashing, 138, 192–205, 238
    collision, 140, 195
    density, 141
    digit-analysis, 192
    division, 140, 193
    extendible, 134, 265–273
    folding, 140, 193
    key-to-address, 140, 192
    linear quotient, 199
    loading factor, 141, 200
    midsquare, 140
    minimal perfect, 204
    offset, 196
    open, 140
    perfect, 201–204
    primary clustering, 196
    pseudorandom, 194
    random, 140
    search, 199
    secondary clustering, 197
    symbol table, 236–240
    synonym, 140
head, 52
    disk, 127
    read/write, 123
HEAD pointer, 63
heap, 149
heap sort, 149

HEAPSORT algorithm, 151
hierarchy, 5
home address, 127
HSORT program, 152
Huffman tree, 163, 176

IBG, 124, 125
implementation, of stack, 80
incidence matrix, 113, 116
indegree, 87
index, 35, 134
indexed sequential file, 135
indexing, string, 28
infix, 242, 244
inorder scan (LNR), 100
input-restricted deque, 81
input/output, *see* I/O
insert, 9, 12, 36
    average number moves, 46
    dense list, 39
    deque, 81
    rate, queue, 70
    stack, 73
    string, 28
    threaded-tree, 105
    tree, 93
INSERT_CIRC_Q, 69
INSERTQ, 64
INSERT_QUEUE, 65
INSERT_THREAD, 106
instantiation, 308
integer, 306
intensity, queue traffic, 70
inter-block gap, 124
internal fragmentation, 208
internal sort, 142
INTERP, 245
interpreter, 245
inter-record gap, 124, 132
I/O, 128
    buffered, 247
I/O buffer, 162, 246
I/O processing, 246–248
IRG, 124, 132

key, 131
    multiple, 131
    single, 131

key sort (tag sort), 166
key-to-address transformation, 140,
    192. *See also* hashing
Kleinrock, Leonard, 70
Knuth, D. E., 201, 210
Königsberg, Bridges of, 120

last in, first out, *see* LIFO
leaf, 88
leaf page, 265
length
    average queue, 70
    average search, 178, 179, 200
    average string, 162
    fixed, 31
    graph path, 113
    string, 28
length of vector, 41
level
    dynamic memory management,
        210
    tree, 87
LIFO, 72
linear list, 58
linear search, 180
link, 34
linked list, 49–56, 61, 226, 289–291
    implementation of stack, 80
    multi-, 56–60
    storage of tree, 95
LINK_INSERT algorithm, 52
LISP, 34
list, 3, 21, 34–60, 286–295
    allocated/free-space, 31, 34
    attribute, 236
    available-space, 213
    circular, 58, 63, 68
    delete, 36, 39
    dense, 34, 61
    dense search, 37
    dense update, 38
    doubly linked, 58, 81
    free-space, 31
    insert, 39
    linear, 58
    linked, 49–56, 61, 226, 289–291
    linked implementation of stack, 80
    multilinked, 56–60, 90

overflow, 46
    singly linked, 81
LNR_DELETE, 188
LNR order, 99, 186
LNR_SCAN, 102
LNR thread, 104
loading factor, 141, 200
load-point marker, 125
local variable, 308
location
    of an array element, 43
    of a three-dimensional element, 43
logical record, 126
Lorin, H., 172
LQHASH, 199
LRN order, 99
Łukasiewicz notation, *see* Polish
    notation

magnetic disk, 126
magnetic drum, 126
magnetic tape, 5, 122–126
    nine-track, 123
    seven-track, 123
main memory access time, 128
main storage, 122
management
    free-space, 81
    memory, 81, 206–223
    storage, 24
map, 7, 85
    road, 110
mark
    end-of-file, 124
    end-of-reel, 125
    load-point, 125
    tape, 124
matrix, 4, 42
    adjacency, 252
    connection, 113
    incidence, 113, 116
    symmetric, 113
memory
    access time, 128
    auxiliary, 122
    fragmentation, 208
    garbage-collection, 209, 235

memory (*continued*)
    main, 122
    secondary, 122
memory management, 81, 206–223
    APL, 212
    best-fit, 209
    Buddy system, 214–217
    dynamic, 210
    Fibonacci, 217–219
    first-fit, 209
    optimizing, 219
merge, 132, 147, 160
    balanced, 167
    cascade, 167
    c-way, 164
    minimal tree, 163
    optimal tree, 163
    order, 161
    three-way, 164
    two-way, 161
MERGE algorithm, 147
midsquare, 140
minimal merge tree, 163
minimal perfect hashing function, 204
minimal spanning tree, 114, 115
MINSPAN algorithm, 115
mode, plotting, 228
moves, average number, 46
multilinked list, 56–60, 90
multiple key file, 131
multiple-path structure, 7

nesting, 310
network, 85, 114
nickname, 35
nil pointer, 50
nine-track tape, 123
NLR order, 99
node, 84, 111
    predecessor, 86
    successor, 86
    terminal, 87
nondirected graph, 84, 111
NRL order, 99
null string, 21
number of paths, 113
numbers, generator of random, 197

odd parity, 123
offset, 196
one-dimensional array, 42
open hash, 140
operating system, 14, 24
operation
    read/write, 128
    string, 27–30
optimal merge tree, 163
optimizing memory management, 219
order
    column-major, 42
    merge, 161
    row-major, 42
organization, tree file, 136
oscillating sort, 172
outdegree, 87, 90
output, *see* I/O
output queue, 61
output-restricted deque, 81
overflow, 46, 235
    deque, 81
    disk, 128
    queue, 63, 65, 70
    stack, 79

page, 265
    depth of, 268
parameter, formal, 309
parent pointer, 92
parity
    even, 123
    odd, 123
parsing, 241–246
Pascal, 21, 26, 42, 49, 66, 92, 115, 203,
    206, 274, 284–285, 302–315
    string functions, 284
pass by reference, 310
pass by variable, 310
path, 85, 111
    alternate, 85
    Hamiltonian, 111
    length, 113
    shortest, 111, 252
    simple, 111
paths, number of, 113
pattern matching, string, 28

peer sort, 166
perfect hashing, 201–204
    minimal, 204
    quotient-reduction, 202
    remainder-reduction, 202
peripheral device, 122
physical record, 126
PL/I, 10, 24, 42, 49, 206, 233, 236, 274–279, 286–295
    ALLOCATE, 288–289
    BASED variables, 288–289
    string functions, 275
plotting mode, 228
pointer
    backward, 58
    forward, 58
    head, 52, 63
    nil, 50
    parent, 92
    tail, 52, 63
Polish notation (NLR), 101, 242
polyphase sort, 169
POP algorithm, 79
pop stack, 74, 77
postfix, 242
postorder scan (LRN), 100
precedence table, 243
predecessor node, 86
preorder scan (NLR), 100
primary clustering, 196
primes, 196–198
printer queue, 62
procedure, 303
program
    HSORT, 152
    QUEUES, 66
program preparation, 224
pseudokey, 267
pseudorandom hashing, 194
pull stack, 77
PUSH algorithm, 79
push down
    stack, 78
    store, 78
push stack, 73, 77

QSORT, 157
quadratic quotient search, 198

queue, 61–72, 81
    deletion rate, 70
    insertion rate, 70
    length average, 70
    output, 61
    overflow, 63, 65, 70
    printer, 62
    traffic intensity, 70
    underflow, 65
QUEUES program, 66
quickersort, 153–158
quotient-reduction perfect hashing, 202

radix, binary, 145
radix sort, 142, 145
radix-sort algorithm, 145
random, 140
random access file, 131
randomizing, 138
random-number generator, 197
read/write head, 123
read/write operation, 128
real, 306
record, 5, 21, 34, 123, 130, 311
    fixed-length, 21, 22
    logical, 126
    physical, 126
    variable-length, 21, 23
recursion, 102, 248–250
reel, tape, 123
reference, pass by, 310
reference counter, 211
register, display, 229
remainder-reduction perfect hashing, 202
replacement, string, 28
replacement selection, 162
replacement symbol, 38
representation of a graph, 112
reserved word, 203, 303
ring, 58
RLN order, 99
RNL order, 99
road map, 7, 85, 110
root, 86
rotational delay, 128
row-major order, 42
run, 161

scan
    data structure, 178
    graph, 117
    inorder (LNR), 100
    postorder (LRN), 100
    preorder (NLR), 100
    tree, 97, 99
SCAN algorithm, 276
scanning arithmetic expression, 74
SCAN_THREAD, 105
scatter storage, 140
search, 12, 36, 128, 178–205
    average length, 178, 179, 200
    binary, 181–183
    binary tree, 186, 291–293
    block, 183–186
    B-tree, 257
    dense-list algorithm, 37
    effort, 180
    hash, 199
    linear, 180
    quadratic quotient, 198
    sequential, 180
    summary, 190
    table, 180
secondary clustering, 197
seek, 128
    delay, 128
segment, string, 166
selection sort, 143
separation, 29
sequential, indexed, 135
sequential access, 126
sequential-access file, 131
sequential search, 180
set, 311, 312
    successor, 86
seven-track tape, 123
Shell sort, 170
shortest path, 111, 252
side effect, 310
simple path, 111
simulation, 273
single key file, 131
single-path structure, 7
singly linked list, 81
SNOBOL, 274, 279–284, 296–301
    string functions, 280

sort, 142–177
    balanced, 167
    binary radix, 145
    binary tree, 90
    bubble, 142, 146
    cascade merge, 167
    comparative, 142
    disk, 174
    distributive, 142
    effort, 143
    external, 142, 160–177
    heap, 149
    internal, 142
    key, 166
    oscillating, 172
    peer, 166
    polyphase, 169
    quicker, 153–158
    radix, 142, 145
    replacement selection, 162
    run, 161
    selection, 143
    string, 161, 162
    summary, 158
    tag, 166
    tree, 149
sort/merge, 160
space versus time, 8
spanning tree, minimal, 114, 115
sparse array, 225–228
SPATH, 251–252
SPATH_ALL, 253
Sprugnoli, R., 202
stack, 72–80, 81, 102, 118, 242–243, 293
    delete, 74
    execution-time, 73
    insert, 73
    linked-list implementation, 80
    overflow, 79
    pop, 74, 77
    pulling, 77
    push, 73, 77, 79
    push-down, 78
    putting, 77
    top of, 72
statement, compound, 304
static structure, 11

storage, 1. *See also* memory
    auxiliary, 122
    contiguous, 34
    of graphs, 111
    main, 122
    scatter, 140
storage management, 24, 81
store, push-down, 78
storing trees, 90–97
string, 15–33, 274–285
    average length, 162
    concatenation, 28
    deletion, 28
    empty, 21
    indexing, 28
    insertion, 28
    length of, 28
    null, 21
    operations, 27–30
    pattern matching, 28
    replacement, 28
    segment of, 166
    sort, 161, 162
structure, 3
    data, 306
    dynamic, 11
    multiple-path, 7
    single-path, 7
    static, 11
    user-defined, 299, 314
structured data types, 311
subfield, 34
subgraph, 86
subrange, 306
subscripted variable, 1
subscripts, 21
substring (copy), 29
subtree, 86
successor node, 86
successor set, 86
summary
    search, 190
    sort, 158
superqueue, 81
superstack, 81
symbol, replacement (←), 38
symbol table, 236–240
SYM_HASH, 237

symmetric matrix, 113
synonym, 140
syntax analyzer, 242
system
    dynamic memory management, 210
    file, 136
    operating, 14, 24

table, 1, 4, 298
    search, 180
    symbol, 236–240
tag sort, 166
tail, 52
TAIL pointer, 63
tape, 5, 122–126
    density, 125
    drive, 123
    end-of-file, 124
    end-of-reel, 125
    file, 131
    load-point, 125
    mark, 124
    nine-track, 123
    read/write head, 123
    reel, 123
    seven-track, 123
    update, 132
telephone directory, 35
terminal, 30
    CRT, 228
terminal node, 87
text, 21
text editing, 30–33, 232–235
thread, LNR, 104
threaded tree, 105
    insert, 105
three-dimensional array, 42
    location of, 43
three-way merge, 164
tic-tac-toe, 6
time, access memory, 128
token, 241
token codes, 245
top of stack, 72
track, 134
    address, 127
    disk, 127

track (*continued*)
    nine, 123
    seven, 123
traffic intensity, queue, 70
transducer, 242
tree, 5, 84, 86
    ancestor, 88
    AVL, 190
    balanced, 190
    binary, 90
    binary search, 186, 291–293
    binary sort, 96
    branch, 88
    B-tree, 134, 254–264
    B$^+$-tree, 261–264
    delete, 107
    descendant, 88
    directed, 86, 91
    directory, 136
    filial set, 88, 164
    game, 6
    Huffman, 163, 176
    indegree, 87
    insert, 93
    leaf, 88
    level, 87
    minimal merge, 163
    minimal spanning, 114, 115
    optimal merge, 163
    organization file, 136
    outdegree, 87, 90
    parent pointer, 92
    predecessor node, 86
    root, 86
    scan, 97, 99
    sort, 149
    spanning, 114, 115
    storage, dense list, 94–95
    storage, linked list, 95
    storing, 90–97
    subtree, 86
    successor node, 86
    thread, 104
    threaded, 105
    threaded insert, 105
    undirected, 86
two-dimensional array, 4, 42
two-way merge, 161
type, 306

underflow, queue, 65
undirected tree, 86
update, 9, 36, 61
    dense list algorithm, 38
    file, 132, 134
    tape, 132
user file, 136
user-defined structures, 299, 314
    Pascal, 314–315
    SNOBOL, 299–301

variable
    global, 308
    local, 308
    pass by, 310
    PL/I based, 288–289
    subscripted, 1
variable-length records, 21, 23
variable-size block, 208
vector, 1, 34, 41
    dope, 234, 235, 239, 240–241
von Neumann, John, 1

weighted digraph, 110
weighted graph, 110, 114
word, reserved, 203, 303